The UK Mathematics Trust

Yearbook

2014 – 2015

This book contains an account of UKMT activities from 1st September 2014 to 31st August 2015. It contains all question papers, solutions and results as well as a variety of other information.

Published by the United Kingdom Mathematics Trust.
School of Mathematics, The University of Leeds, Leeds LS2 9JT
Telephone: 0113 343 2339
E-mail: enquiry@ukmt.org.uk
Website: http://www.ukmt.org.uk

Cover design: – The backdrop is a Penrose tiling whose complexity reflects the activities of the UKMT.

The photographs are

Front Cover:

IMC and JMC winners, St Aloysius' College, Glasgow

Back Cover:

UK team at the International Mathematical Olympiad 2015

National Mathematics Summer School, Oxford, August 2014

ISBN 978-1-906001-26-1

Printed and bound in Great Britain by
H. Charlesworth & Co. Ltd, Wakefield

Contents

The Senior Mathematical Challenge and British Mathematical Olympiads

Olympiad Training and Overseas Competitions

56th International Mathematical Olympiad, Chiang Mai, Thailand

UKMT Mentoring Schemes

Team Maths Challenge

Senior Team Maths Challenge

Other aspects of the UKMT

Other similar bodies overseas

Lists of volunteers involved in the UKMT's activities

UKMT Publications

This is my last opportunity to introduce the UKMT Yearbook, as I will be handing over the chair of the UKMT Council to Chris Budd after the AGM March 2016. We were all delighted (if not surprised) when Chris received an OBE in the Queen's Birthday Honours this year for outstanding services to science and maths education.

I have thoroughly enjoyed my time chairing the Council since April 2010, and I look forward to remaining involved with the Trust in different ways in the future. In my opinion it is a wonderful institution … thanks to its combination of fantastic volunteers and dedicated staff doing truly excellent work to inspire hundreds of thousands of UK school children to enjoy and excel at mathematics.

The Trust's core activity is of course the running of the Maths Challenges, and I am very happy to be able to report that the total number of entries to the Challenges was the highest ever in 2014/15, exceeding the previous year's record total. Entry numbers for the Junior Challenge rose from 290,820 in 2013/14 to 293,940 and the number of participating schools also increased from 3,909 to 3,970. For the Intermediate Challenge numbers of entries and participating schools increased from 254,130 entries and 3,147 schools in 2013/14 to 259,480 entries and 3,199 schools in 2014/15. Likewise numbers for the Senior Challenge rose again from 104,360 entries and 2,113 schools in 2013/14 to 109,660 entries and 2,206 schools in 2014/15. In addition entries for the Team Challenge and for the Senior Team Challenge (the latter run in collaboration with the Further Maths Support Programme) both rose again slightly this year to 1,754 and 1,169 teams respectively.

Our congratulations go to the UK's IMO team, who returned from this year's International Olympiad competition in Thailand with four silver medals, one bronze medal and an honourable mention, finishing 22nd out of 104 teams.

Congratulations will be due next year to the Trust itself, whose 20th anniversary will happen in 2016. A Volunteers' Meeting including a celebration of the anniversary is being planned. I am sure that the Trust will continue to flourish in the coming years, and indeed will go from strength to strength under the excellent leadership of its Director Rachel Greenhalgh and my successor Chris Budd!

Frances Kirwan
Balliol College, Oxford
August 2015

Introduction

Foundation of the Trust

National mathematics competitions have existed in the UK for several decades. Up until 1987 the total annual participation was something like 8,000. Then there was an enormous growth, from 24,000 in 1988 to around a quarter of a million in 1995 – without doubt due to the drive, energy and leadership of Dr Tony Gardiner. By the end of this period there were some nine or ten competitions for United Kingdom schools and their students organised by three different bodies: the British Mathematical Olympiad Committee, the National Committee for Mathematical Contests and the UK Mathematics Foundation. During 1995 discussions took place between interested parties which led to agreement to seek a way of setting up a single body to continue and develop these competitions and related activities. This led to the formation of the United Kingdom Mathematics Trust, which was incorporated as a company limited by guarantee in October 1996 and registered with the Charity Commission.

Throughout its existence, the UKMT has continued to nurture and expand the number of competitions. As a result, over six hundred thousand students throughout the UK now participate in the challenges alone, and their teachers (as well as others) not only provide much valued help and encouragement, but also take advantage of the support offered to them by the Trust.

The Royal Institution of Great Britain is the Trust's Patron, and it and the Mathematical Association are Participating Bodies. The Association of Teachers of Mathematics, the Edinburgh Mathematical Society, the Institute of Mathematics and Its Applications, the London Mathematical Society and the Royal Society are all Supporting Bodies.

Aims and Activities of the Trust

According to its constitution, the Trust has a very wide brief, namely "to advance the education of children and young people in mathematics". To attain this, it is empowered to engage in activities ranging from teaching to publishing and lobbying. But its focal point is the organisation of mathematical competitions, from popular mass "challenges" to the selection and training of the British team for the annual International Mathematical Olympiad (IMO).

There are three main challenges, the UK Junior, Intermediate and Senior Mathematical Challenges. The number of challenge entries in 2014-2015 totalled 663,080: once again, a pleasing increase in entry numbers year on year. The challenges were organised by the Challenges Subtrust (CS).

The Challenges are open to all pupils of the appropriate age. Certificates are awarded for the best performances and the most successful participants are encouraged to enter follow-up competitions.

At the junior and intermediate levels, we increased the number of pupils entering follow-up competitions from a total of around 11,000 to a total of around 16,600. This significant increase was due in large part to the introduction of a Junior Kangaroo. The follow-up rounds now consist of the Junior Olympiad and Kangaroo, and a suite of papers forming the Intermediate Olympiad and Kangaroo under the auspices of the Challenges Subtrust.

The British Mathematical Olympiad Committee Subtrust (BMOS) organises two rounds of the British Mathematical Olympiad. Usually about 800 students who have distinguished themselves in the Senior Mathematical Challenge are invited to enter Round 1, leading to about 100 in Round 2. From the latter, around twenty are invited to a training weekend at Trinity College, Cambridge. Additionally, an elite squad, identified largely by performances in the UKMT competitions, is trained at camps and by correspondence courses throughout the year. The UK team is then selected for the annual International Mathematical Olympiad (IMO) which usually takes place in July. Recent IMOs were held as follows: UK (2002), Japan (2003), Athens (2004), Mexico (2005), Slovenia (2006), Vietnam (2007), Madrid (2008), Bremen (2009), Kazakhstan (2010), Amsterdam (2011), Argentina (2012), Colombia (2013), South Africa (2014) and Thailand in 2015. The BMOS also runs a mentoring scheme for high achievers at senior, intermediate and junior levels.

There is a Kangaroo follow-on round at the senior level as well, and over 3,000 pupils are invited to participate each year.

Structure and Membership of the Trust

The governing body of the Trust is its Council. The events have been organised by three Subtrusts who report directly to the Council. The work of the Trust in setting question papers, marking scripts, monitoring competitions, mentoring students and helping in many other ways depends critically on a host of volunteers. A complete list of members of the Trust, its Subtrusts and other volunteers appears at the end of this publication.

Challenges Office Staff

Rachel Greenhalgh continues in her role as Director of the Trust, ably supported by the Maths Challenges Office staff of Nicky Bray, Janet Clark, Gerard Cummings, Heather Macklin, Shona Raffle-Edwards and Jo Williams. Beverley Detoeuf continues as Packing Office Manager and leads the packing and processing team of Terri-J de Nobrega, Rachael Raby-Cox, Stewart Ramsay, Alison Steggall and Tabitha Taylor, ably assisted by Mary Roberts, Packing Office Supervisor.

An outline of the events

This is a brief description of the challenges, their follow-up competitions and other activities. Much fuller information can be found later in the book.

Junior competitions

The UK Junior Mathematical Challenge, typically held on the last Thursday in April, is a one hour, 25 question, multiple choice paper for pupils up to and including:

Y8 in England and Wales; S2 in Scotland, and Y9 in Northern Ireland.

Pupils enter their personal details and answers on a special answer sheet for machine reading. The questions are set so that the first 15 should be accessible to all participants whereas the remaining 10 are more testing.

Five marks are awarded for each correct answer to the first 15 questions and six marks are awarded for each correct answer to the rest. Each incorrect answer to questions 16–20 loses 1 mark and each incorrect answer to questions 21–25 loses 2 marks. Penalty marking is used to discourage guessing.

Certificates are awarded on a proportional basis:– Gold about 6%, Silver about 14% and Bronze about 20% of all entrants. Each centre also receives one 'Best in School Certificate'. A 'Best in Year Certificate' is awarded to the highest scoring candidate in each year group, in each school.

Until 2014, the Junior Mathematical Olympiad was the only follow-up competition to the JMC. In 2015 we were delighted to introduce the Junior Kangaroo so that all our individual Challenges now have both Olympiad and Kangaroo follow-up rounds. The Junior Mathematical Olympiad and Kangaroo (JMOK) is held around 6 weeks after the JMC. Between 1,000-1,200 high scorers in the JMC are invited to take part in the Olympiad; the next 5,000 or so are invited to take part in the Kangaroo.

The Olympiad is a two-hour paper with two sections. Section A contains ten questions and pupils are required to give the answer only. Section B contains six questions for which full written answers are required. It is made clear to candidates that they are not expected to complete all of Section B and that little credit will be given to fragmentary answers. Gold, silver and bronze medals are awarded to very good candidates. In 2015, a total of 215 medals was awarded. The top 25% candidates got Certificates of Distinction. Most of the rest receive a Merit and of the rest, those who had qualified for the Olympiad automatically via the JMC received a Certificate of Qualification. In addition, the top 50 students were given book prizes.

The Junior Mathematical Kangaroo is a one-hour multiple-choice paper, with 25 questions (like the JMC, but more challenging!). Certificates of Merit are awarded to the top 25% and certificates of Qualification to everyone else who takes part.

Intermediate competitions

The UK Intermediate Mathematical Challenge is organised in a very similar way to the Junior Challenge. One difference is that the age range goes up to Y11 in England and Wales, to S4 in Scotland and Y12 in Northern Ireland. The other difference is the timing; the IMC is held on the first Thursday in February. All other arrangements are as in the JMC.

There are five follow-up competitions under the overall title 'Intermediate Mathematical Olympiad and Kangaroo' (IMOK). Between 400 and 550 in each of Years 9, 10 and 11 (English style) sit an Olympiad paper (Cayley, Hamilton and Maclaurin respectively). In 2015, each of these was a two-hour paper and contained six questions all requiring full written solutions. A total of around 9000 pupils from the three year groups took part in a Kangaroo paper. In the European Kangaroo papers, which last an hour, there are 25 multiple-choice questions. The last ten questions are more testing than the first fifteen and correct answers gain six marks as opposed to five. Penalty marking is not applied. The same Kangaroo paper (designated 'Pink') was taken by pupils in Years 10 and 11 and a different one, 'Grey', by pupils in Year 9. In 2015, the Olympiads and Kangaroos were sat on Thursday 19th March. In the Olympiads, the top 25% of candidates got Certificates of Distinction. Most of the rest receive a Merit and of the rest, those who had qualified for the Olympiad automatically via the IMC received a Certificate of Participation. In the Kangaroos, the top 25% got a Merit and the rest a Participation. All Olympiad and Kangaroo candidates received a 'Kangaroo gift'; a specially designed UKMT key fob. In addition, the top 50 students in each year group in the Olympiad papers were given a book. Performance in the Olympiad papers and the IMC was a major factor in determining pupils to be invited to one of the UKMT summer schools early in July.

Senior competitions

In 2014, the UK Senior Mathematical Challenge was held on Thursday 6th November. Like the other Challenges, it is a 25 question, multiple choice paper marked in the same way as the Junior and Intermediate Challenges. However, it lasts 1½ hours. Certificates (including Best in School) are awarded as with the other Challenges. The follow-up competitions are the British Mathematical Olympiads 1 and 2 (organised by the British Mathematical Olympiad Subtrust) and the Senior Kangaroo.

The first Olympiad stage, BMO1, was held on Friday 28th November

2014. About 800 are usually invited to take part. The paper lasted 3½ hours and contained six questions to which full written solutions are required.

About 100 high scorers are then invited to sit BMO2, which was held on Thursday 29th January 2015. It also lasted 3½ hours but contained four, very demanding, questions.

The results of BMO2 are used to select a group of students to attend a Training Session at Trinity College, Cambridge at Easter. As well as being taught more mathematics and trying numerous challenging problems, this group sits a 4½ hour 'mock' Olympiad paper. On the basis of this and all other relevant information, a group of about eight is selected to take part in correspondence courses and assignments which eventually produce the UK Olympiad Team of six to go forward to the International Mathematical Olympiad in July. In 2014, the Senior Kangaroo paper, for pupils who were close to being eligible for BMO1, was held on the same day, with the number of participants staying at around 3,300.

The growth of the Challenges

In the 2005 UKMT Yearbook, we showed the growth of the Challenges since UKMT was established and this has now been updated. The graphs below show two easily identifiable quantities, the number of schools and the number of entries. In each case, the lines, from top to bottom, represent the Junior, Intermediate and Senior Challenges. As those involved in the UKMT firmly believe that the Challenges are a very worthwhile endeavour, we hope that the upward trends are continued.

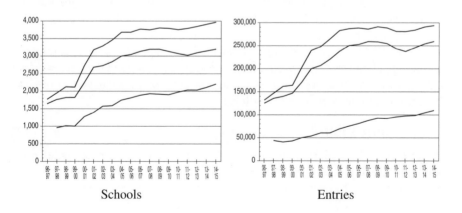

Schools   Entries

Team Maths Challenge and Senior Team Maths Challenge

This event is the successor of the Enterprising Mathematics UK which was run in conjunction with the IMO in 2002. A team consists of four pupils who are in Year 9 (English style) or below with at most two members being in Year 9. In 2015, over 1600 teams took part in Regional Finals and 88 schools competed in the National Final held in the grand surroundings of the Lindley Hall, part of the prestigious Royal Horticultural Halls in Westminster, London, on Monday 22nd June 2015.

In addition, 1169 schools took part in the Senior Team Maths Challenge which is aimed at students studying maths beyond GCSE. The final, which involved 80 teams, was held at the Lindley Hall, part of the Royal Horticultural Halls, London, on Tuesday 3rd February 2015.

Report from the Director

It gives me great pleasure to read about the Trust's activities over the past year, and I hope you enjoy reading the Yearbook too. Thank you for your help in making 2014/15 another very successful year for the Trust. I am heartened that despite the very challenging economic situation, and the fast pace of change happening in many schools and colleges today, so many of you continue to participate in, support, and benefit from our activities.

The growth in participation in the Maths Challenges this year was particularly pleasing, and we enjoyed seeing and hearing about student success in these via social media (do follow us on Twitter @UKMathsTrust). Success in school mathematics is a great cause for celebration, and it is often overlooked. We are very grateful to the work of teachers who organise the Challenges within schools, and the support of their colleagues, in helping us develop our students' critical thinking. Do let us know if you feel there are things we should be doing to further support mathematics education.

This year saw the introduction of a new follow-on competition, the Junior Kangaroo. This provided an opportunity for further recognition for those students who had scored well in the JMC. If you have not seen this paper, it is featured in this Yearbook, so why not have a go?

Our thanks go to our key supporters and donors, in particular to the Institute and Faculty of Actuaries which continues to support the Challenges, and to Oxford Asset Management which supports our mentoring schemes and IMO team. We were also fortunate and grateful to receive donations, large and small, via www.donate.ukmt.org.uk.

My final thanks are saved for our wonderful group of volunteers and staff who I work with throughout the year and without whom we could not run any of our activities or events! Our volunteers do so much on our behalf, and I would like to publicly thank them all for their commitment and enthusiasm throughout the year.

Rachel Greenhalgh
director@ukmt.org.uk

Institute
and Faculty
of Actuaries

Profile

The Institute and Faculty of Actuaries (IFoA) is the UK's only chartered professional body dedicated to educating, developing and regulating actuaries based both in the UK and internationally.

What is an actuary?

Actuaries are experts in risk management. They use their mathematical skills to help measure the probability and risk of future events. This information is useful to many industries, including healthcare, pensions, insurance, banking and investments, where a single decision can have a major financial impact.

It is a global profession with internationally-recognised qualifications. It is also very highly regarded, in the way that medicine and law are, and an actuarial career can be one of the most diverse, exciting and rewarding in the world. In fact, due to the difficult exams and the expertise required, being an actuary carries quite a reputation.

How do I become an actuary?

To qualify as an actuary you need to complete the IFoA's exams. Most actuarial trainees take the exams whilst working for an actuarial employer. Exemptions from some of the exams may be awarded to students who have studied to an appropriate standard in a relevant degree, or have studied actuarial science at Postgraduate level. Qualification typically takes three-six years. Those on a graduate actuarial trainee programme can expect to earn £29,000-£35,000 a year. This will increase to well over £100,000 as you gain more experience and seniority.

International outlook

The IFoA qualification is already highly valued throughout the world, with 42% of its members based outside the UK. Mutual recognition agreements with other international actuarial bodies facilitate the ability for actuaries to move and work in other parts of the world and create a truly global profession.

For more information on the qualifications and career path visit our website – http://www.actuaries.org.uk/becoming-actuary or join us on Facebook – www.be-an-actuary.co.uk

The Junior Mathematical Challenge and Olympiad

The Junior Mathematical Challenge was held on Thursday 30th April 2015 and over 254,000 pupils took part. Approximately 1000 pupils were invited to take part in the Junior Mathematical Olympiad, and a further 5000 to take part in the Junior Mathematical Kangaroo, both of which were held on Tuesday 9th June. In the following pages, we shall show the question paper and solutions leaflet for both the JMC and JMO.

We start with the JMC paper, the front of which is shown below in a slightly reduced format.

UK JUNIOR MATHEMATICAL CHALLENGE

THURSDAY 30th APRIL 2015

Organised by the **United Kingdom Mathematics Trust**
from the School of Mathematics, University of Leeds

Institute
and Faculty
of Actuaries

RULES AND GUIDELINES (to be read before starting)

1. Do not open the paper until the Invigilator tells you to do so.

2. Time allowed: **1 hour**.
   No answers, or personal details, may be entered after the allowed hour is over.

3. The use of rough paper is allowed; **calculators** and measuring instruments are **forbidden**.

4. Candidates in England and Wales must be in School Year 8 or below.
   Candidates in Scotland must be in S2 or below.
   Candidates in Northern Ireland must be in School Year 9 or below.

5. **Use B or HB non-propelling pencil only.** Mark *at most one* of the options A, B, C, D, E on the Answer Sheet for each question. Do not mark more than one option.

6. *Do not expect to finish the whole paper in 1 hour.* Concentrate first on Questions 1-15. When you have checked your answers to these, have a go at some of the later questions.

7. Five marks are awarded for each correct answer to Questions 1-15.
   Six marks are awarded for each correct answer to Questions 16-25.
   Each incorrect answer to Questions 16-20 loses 1 mark.
   Each incorrect answer to Questions 21-25 loses 2 marks.

8. Your Answer Sheet will be read only by a *dumb machine*. **Do not write or doodle on the sheet except to mark your chosen options.** The machine 'sees' all black pencil markings even if they are in the wrong places. If you mark the sheet in the wrong place, or leave bits of rubber stuck to the page, the machine will 'see' a mark and interpret this mark in its own way.

9. The questions on this paper challenge you to **think**, not to guess. You get more marks, and more satisfaction, by doing one question carefully than by guessing lots of answers. The UK JMC is about solving interesting problems, not about lucky guessing.

The UKMT is a registered charity
http://www.ukmt.org.uk

1. Which of the following calculations gives the largest answer?

   A $1-2+3+4$ B $1+2-3+4$ C $1+2+3-4$ D $1+2-3-4$ E $1-2-3+4$

2. It has just turned 22:22. How many minutes are there until midnight?

   A 178   B 138   C 128   D 108   E 98

3. What is the value of $\dfrac{12\,345}{1+2+3+4+5}$?

   A 1   B 8   C 678   D 823   E 12 359

4. In this partly completed pyramid, each rectangle is to be filled with the sum of the two numbers in the rectangles immediately below it.

   What number should replace x?

   A 3   B 4   C 5   D 7   E 12

5. The difference between $\dfrac{1}{3}$ of a certain number and $\dfrac{1}{4}$ of the same number is 3. What is that number?

   A 24   B 36   C 48   D 60   E 72

6. What is the value of x in this triangle?

   A 45   B 50   C 55   D 60   E 65

7. The result of the calculation 123 456 789 × 8 is almost the same as 987 654 321 except that two of the digits are in a different order. What is the sum of these two digits?

   A 3   B 7   C 9   D 15   E 17

8. Which of the following has the same remainder when it is divided by 2 as when it is divided by 3?

   A 3   B 5   C 7   D 9   E 11

9. According to a newspaper report, "A 63-year-old man has rowed around the world without leaving his living room." He clocked up 25 048 miles on a rowing machine that he received for his 50th birthday.

   Roughly how many miles per year has he rowed since he was given the machine?

   A 200   B 500   C 1000   D 2000   E 4000

10. In the expression 1 □ 2 □ 3 □ 4   each □ is to be replaced by either + or ×.

    What is the largest value of all the expressions that can be obtained in this way?

    A 10   B 14   C 15   D 24   E 25

11. What is the smallest prime number that is the sum of three different prime numbers?

    A 11   B 15   C 17   D 19   E 23

12. A fish weighs the total of 2 kg plus a third of its own weight. What is the weight of the fish in kg?

    A $2\frac{1}{3}$ B 3   C 4   D 6   E 8

13. In the figure shown, each line joining two numbers is to be labelled with the sum of the two numbers that are at its end points.

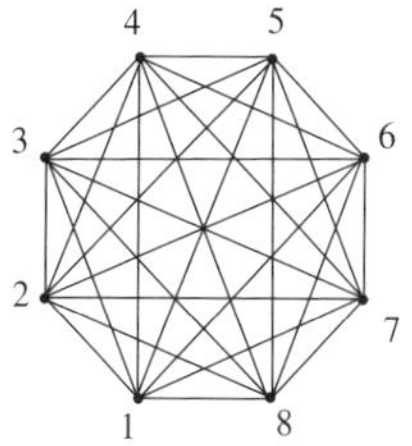

    How many of these labels are multiples of 3?

    A 10   B 9   C 8   D 7   E 6

14. Digits on a calculator are represented by a number of horizontal and vertical illuminated bars. The digits and the bars which represent them are shown in the diagram.

    How many digits are both prime and represented by a prime number of illuminated bars?

    A 0   B 1   C 2   D 3   E 4

15. Which of the following is divisible by all of the integers from 1 to 10 inclusive?

    A 23×34 B 34×45 C 45×56 D 56×67 E 67×78

16. The diagram shows a square inside an equilateral triangle. What is the value of $x + y$?

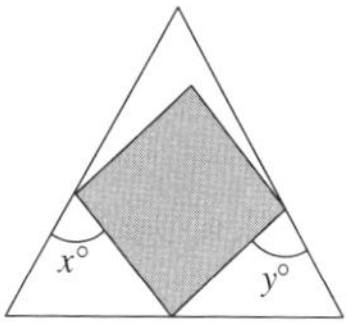

    A 105   B 120   C 135   D 150   E 165

17.

    Knave of Hearts: "I stole the tarts."
    Knave of Clubs: "The Knave of Hearts is lying."
    Knave of Diamonds: "The Knave of Clubs is lying."
    Knave of Spades: "The Knave of Diamonds is lying."

    How many of the four Knaves were telling the truth?

    A 1   B 2   C 3   D 4   E more information needed

18. Each of the fractions $\frac{2637}{18\,459}$ and $\frac{5274}{36\,918}$ uses the digits 1 to 9 exactly once. The first fraction simplifies to $\frac{1}{7}$. What is the simplified form of the second fraction?

    A $\frac{1}{8}$ B $\frac{1}{7}$ C $\frac{5}{34}$ D $\frac{9}{61}$ E $\frac{2}{7}$

19. One of the following cubes is the smallest cube that can be written as the sum of three positive cubes. Which is it?

   A 27   B 64   C 125   D 216   E 512

20. The diagram shows a pyramid made up of 30 cubes, each measuring $1\,m \times 1\,m \times 1\,m$.
   What is the total surface area of the whole pyramid (including its base)?

   A $30\,m^2$ B $62\,m^2$ C $72\,m^2$ D $152\,m^2$ E $180\,m^2$

21. Gill is now 27 and has moved into a new flat. She has four pictures to hang in a horizontal row on a wall which is 4800 mm wide. The pictures are identical in size and are 420 mm wide. Gill hangs the first two pictures so that one is on the extreme left of the wall and one is on the extreme right of the wall. She wants to hang the remaining two pictures so that all four pictures are equally spaced. How far should Gill place the centre of each of the two remaining pictures from a vertical line down the centre of the wall?

   A 210 mm   B 520 mm   C 730 mm   D 840 mm   E 1040 mm

22. The diagram shows a shaded region inside a regular hexagon. The shaded region is divided into equilateral triangles. What fraction of the area of the hexagon is shaded?

   A $\dfrac{3}{8}$ B $\dfrac{2}{5}$ C $\dfrac{3}{7}$ D $\dfrac{5}{12}$ E $\dfrac{1}{2}$

23. The diagram shows four shaded glass squares, with areas 1 cm², 4 cm², 9 cm² and 16 cm², placed in the corners of a rectangle. The largest square overlaps two others. The area of the region inside the rectangle but not covered by any square (shown unshaded) is 1.5 cm².
   What is the area of the region where squares overlap (shown dark grey)?

   A $2.5cm^2$ B $3cm^2$ C $3.5cm^2$ D $4cm^2$ E $4.5cm^2$

24. A *palindromic number* is a number that reads the same when the order of its digits is reversed. What is the difference between the largest and smallest five-digit palindromic numbers that are both multiples of 45?

   A 9180   B 9090   C 9000   D 8910   E 8190

25. The four straight lines in the diagram are such that $VU = VW$. The sizes of $\angle UXZ$, $\angle VYZ$ and $\angle VZX$ are $x°$, $y°$ and $z°$.
   Which of the following equations gives x in terms of y and z?

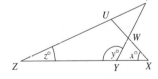

   A $x = y - z$ B $x = 180 - y - z$ C $x = y - \dfrac{z}{2}$

            D $x = y + z - 90$ E $x = \dfrac{y - z}{2}$

The JMC solutions

The usual solutions leaflet was issued.

UK JUNIOR MATHEMATICAL CHALLENGE

THURSDAY 30th APRIL 2015

Organised by the **United Kingdom Mathematics Trust**
from the School of Mathematics, University of Leeds

http://www.ukmt.org.uk

Institute
and Faculty
of Actuaries

SOLUTIONS LEAFLET

This solutions leaflet for the JMC is sent in the hope that it might provide all concerned with some alternative solutions to the ones they have obtained. It is not intended to be definitive. The organisers would be very pleased to receive alternatives created by candidates.

For reasons of space, these solutions are necessarily brief. There are more in-depth, extended solutions available on the UKMT website, which include some exercises for further investigation:

http://www.ukmt.org.uk/

The UKMT is a registered charity

1. **A** The values of the expressions are: A 6, B 4, C 2, D –4, E 0.
(*Alternative method: since every expression contains the integers 1, 2, 3 and 4, the expression which has the largest value is that in which the sum of the integers preceded by a minus sign is smallest. This is expression A.*)

2. **E** At 22:22, there are $60 - 22 = 38$ minutes to 23:00. There are then a further 60 minutes to midnight. So the number of minutes which remain until midnight is $38 + 60 = 98$.

3. **D** The value of $\dfrac{12\,345}{1 + 2 + 3 + 4 + 5} = \dfrac{12\,345}{15} = \dfrac{2469}{3} = 823$.

4. **A** The calculations required to find the value of x are:
$p = 105 - 47 = 58$; $q = p - 31 = 58 - 31 = 27$;
$r = 47 - q = 47 - 27 = 20$;
$s = r - 13 = 20 - 13 = 7$; $t = 13 - 9 = 4$;
$x = s - t = 7 - 4 = 3$.
(*Note that the problem may be solved without finding the values of four of the numbers in the pyramid. Finding these is left as an exercise for the reader.*)

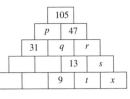

5. **B** Let the required number be x. Then $\frac{x}{3} - \frac{x}{4} = 3$. Multiplying both sides by 12 gives $4x - 3x = 36$. So $x = 36$.

6. **B** The sum of the exterior angles of any polygon is 360°. So $y = 360 - (110 + 120) = 360 - 230 = 130$.
The sum of the angles on a straight line is 180°, so $x = 180 - y = 180 - 130 = 50$.

7. **A** The units digit of $123\,456\,789 \times 8$ is 2, since $9 \times 8 = 72$. So, if the statement in the question is correct then the two digits which are in a different order are 1 and 2, whose sum is 3. As a check, $123\,456\,789 \times 8$ is indeed $987\,654\,312$.

8. **C** All of the options are odd and therefore give a remainder of 1 when divided by 2. Two of the options, 3 and 9, give remainder 0 when divided by 3. Two other options, 5 and 11, give remainder 2 when divided by 3, and 7 is the only option which gives remainder 1 when divided by 3.

9. **D** The man has rowed the equivalent of just over 25 000 miles in approximately 13 years. So the mean number of 'miles' rowed per year is approximately $\dfrac{25\,000}{13} \approx \dfrac{26\,000}{13} = 2000$.

10. **E** If m and n are positive integers, then $mn > m + n$ unless at least one of m or n is equal to 1, or $m = n = 2$. So, to maximise the expression, we need to place multiplication signs between 2 and 3 and between 3 and 4. However, we need to place an addition sign between 1 and 2 because $1 + 2 \times 3 \times 4 = 25$, whereas $1 \times 2 \times 3 \times 4 = 24$.

11. **D** It can be established that 2 is not one of the three primes to be summed since the sum of 2 and two other primes is an even number greater than 2 and therefore not prime. The smallest three odd primes are 3, 5, 7 but these sum to 15 which is not prime. The next smallest sum of three odd primes is $3 + 5 + 11 = 19$, which is prime. So 19 is the smallest prime which is the sum of three different primes.

12. **B** The question tells us that 2 kg is two-thirds of the weight of the fish. So one-third of its weight is 1 kg and therefore its weight is 3 kg.

13.　A　We denote the label joining m and n as $(m + n)$. The labels which are multiples of 3 are $(1 + 2), (1 + 5), (1 + 8), (2 + 4), (2 + 7), (3 + 6), (4 + 5), (4 + 8),$ $(5 + 7), (7 + 8)$. So 10 of the labels are multiples of 3.

14.　E　The primes and the number of illuminated bars which represent them are: $2 \rightarrow 5, 3 \rightarrow 5, 5 \rightarrow 5, 7 \rightarrow 3$. So all four prime digits are represented by a prime number of illuminated bars.

15.　C　Of the options given, 23×34, 56×67 and 67×78 are all not divisible by 5, so may be discounted. Also 34 is not divisible by 4 and 45 is odd, so 34×45 may also be discounted as it is not divisible by 4. The only other option is 45×56. As a product of prime factors, $45 \times 56 = 2^3 \times 3^2 \times 5 \times 7$, so it is clear that it is divisible by all of the integers from 1 to 10 inclusive.

16.　D　The size of each interior angle of an equilateral triangle is 60°. As the sum of the interior angles of a triangle is $180°, x + p + 60 = 180$, so $p = 120 - x$. Similarly, $q = 120 - y$. Each interior angle of a square is a right angle and the sum of the angles on a straight line is $180°$, so $p + q + 90 = 180$. Therefore $120 - x + 120 - y + 90 = 180$, that is $330 - (x + y) = 180$. So $x + y = 330 - 180 = 150$.

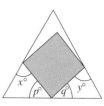

17.　B　If the Knave of Hearts is telling the truth then the Knave of Clubs is lying, which means that the Knave of Diamonds is telling the truth, but the Knave of Spades is lying. Alternatively, if the Knave of Hearts is lying then the Knave of Clubs is telling the truth, which means that the Knave of Diamonds is lying, but the Knave of Spades is telling the truth. In both cases, we can determine that two of the Knaves are lying, although it is not possible to determine which two they are.

18.　B　The fraction $\dfrac{5274}{36\,918} = \dfrac{2637}{18\,459} = \dfrac{1}{7}$, as given in the question.

19.　D　The first six positive cubes are 1, 8, 27, 64, 125, 216. Clearly, 64 cannot be the sum of three positive cubes as the sum of all the positive cubes smaller than 64 is $1 + 8 + 27 = 36$. Similarly, 125 cannot be the sum of three positive cubes as the largest sum of any three positive cubes smaller than 125 is $8 + 27 + 64 = 99$. However, we note that $27 + 64 + 125 = 216$, so 216 is the smallest cube which is the sum of three positive cubes.

20.　C　When the pyramid is viewed from above, it can be seen that the total area of the horizontal part of the surface of the pyramid (excluding its base) is the same as that of a square of side 4 metres, that is 16 m². The area of the base of the pyramid is also 16 m². Finally the total area of the vertical part of the pyramid is equal to $(4 \times 1 + 4 \times 2 + 4 \times 3 + 4 \times 4)$ m² $= 40$ m². So the total surface area of the pyramid is $(16 + 16 + 40)$ m² $= 72$ m².

21.　C　The diagram shows part of the wall of width 4800 mm and the four equally spaced pictures, each of width 420 mm. Let x be the required distance, that is the distance from the centre of each of the two pictures in the middle of the

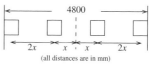

wall to a vertical line down the centre of the wall (marked by a broken line). Then the distance between the centres of any two adjacent pictures is $2x$. Note that the distance between the centres of the two pictures on the extremes of the wall is $(4800 - 2 \times 210)$ mm $= 4380$ mm. Therefore $2x + x + x + 2x = 4380$. So $x = 4380 \div 6 = 730$. Hence the required distance is 730 mm.

22. **E** In the diagram, the shaded small equilateral triangles have been divided into those which lie within the highlighted large equilateral triangle and the twelve small equilateral triangles which lie outside the large triangle.

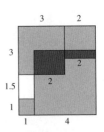

Note that the unshaded star shape in the centre of the large triangle is made up of twelve small equilateral triangles, so the small triangles outside the large triangle could be moved into the large triangle so that the large triangle is shaded completely and the rest of the hexagon is unshaded as in the lower diagram.

The lower diagram shows that the hexagon may be divided into six congruent triangles, three of which are shaded and three of which are unshaded. So the required fraction is $\frac{1}{2}$.

23. **D** The diagram shows some of the lengths of sides which may be deduced from the information given in the question. Note that the rectangle measures 5 cm by 5.5 cm. The sum of the areas of the four glass squares is $(1 + 4 + 9 + 16)$ cm^2 = 30 cm^2. However, the total region of the rectangle occupied by the four squares is equal to $(5 \times 5.5 - 1.5)$ cm^2 = 26 cm^2. So the area of the overlap is $(30 - 26)$ cm^2 = 4 cm^2.

24. **B** For a number to be a multiple of 45 it must be a multiple of 5 and also of 9. In order to be a multiple of 5, a number's units digit must be 0 or 5. However, the units digit of a palindromic number cannot be 0, so it may be deduced that any palindromic number which is a multiple of 45 both starts and ends in the digit 5. In order to make the desired number as large as possible, its second digit should be 9 and for it to be as small as possible its second digit should be 0. So, if possible, the numbers required are of the form '59x95' and '50y05'. In addition, both numbers are to be multiples of 9 which means the sum of the digits of both must be a multiple of 9. For this to be the case, x = 8 and y = 8, giving digit sums of 36 and 18 respectively. So the two required palindromic numbers are 59895 and 50805. Their difference is 9090.

25. **E** The exterior angle of a triangle is equal to the sum of its two interior and opposite angles.
Applying this theorem to triangle UZX:
$\angle VUW = z° + x°$.
Similarly, in triangle WYX: $y° = \angle XWY + x°$, so $\angle XWY = y° - x°$.
As $VU = VW$, $\angle VUW = \angle VWU$ and also $\angle VWU = \angle XWY$ because they are vertically opposite angles. Therefore $\angle VUW = \angle XWY$. So $z° + x° = y° - x°$ and hence $x = \frac{1}{2}(y - z)$.

The JMC answers

The table below shows the proportion of pupils' choices. The correct answer is shown in bold. [The percentages are rounded to the nearest whole number.]

Qn	A	B	C	D	E	Blank
1	**90**	3	4	0	2	1
2	3	8	1	4	**83**	1
3	2	3	15	**73**	3	4
4	**75**	7	5	6	4	3
5	9	**73**	7	3	3	4
6	5	**82**	2	7	2	2
7	**54**	6	16	11	6	6
8	3	10	**71**	3	8	4
9	6	6	10	**71**	4	3
10	4	2	4	69	**17**	3
11	21	39	12	**13**	10	5
12	54	**29**	4	5	2	5
13	**19**	14	17	15	24	8
14	5	7	10	22	**51**	5
15	7	19	**40**	9	10	15
16	5	18	7	**22**	5	42
17	17	**31**	5	1	24	20
18	4	**28**	5	4	11	47
19	10	9	6	**14**	2	58
20	18	7	**15**	4	5	50
21	5	7	**8**	7	11	62
22	3	4	6	7	**15**	65
23	3	4	7	**10**	7	69
24	3	**7**	4	4	3	78
25	5	7	3	4	**4**	76

JMC 2015: Some comments on the pupils' choices of answers as expressed in the feedback letter to schools

It is pleasing to report that the average score of around 48 was slightly higher than last year, and that each of the first nine questions were answered correctly by over half the pupils and, in several cases, by more than three-quarters of them.

However, it looks as though after this promising start many pupils turned their brains off! The responses to questions 10, 11 and 12 were particularly disappointing. You can see from your results sheets how your pupils fared on these questions. We hope you will find the time to discuss the questions with your pupils, or, at least, to direct their attention to the extended solutions on our website.

The vast majority of pupils, faced with question 10, seem to have assumed, without much thought, that the largest value is obtained by using nothing but multiplication signs. A little thought would show that multiplying by 1 does not achieve anything, and a little more thought would show that $1 + 2 \times 3 \times 4$ is larger than $1 \times 2 \times 3 \times 4$.

Question 11 is harder than the earlier questions, and it looks as though many pupils did not give it the careful thought that it needs. One-fifth of the pupils chose 11 as their answer. This is the smallest prime among the given options, but fails to be the sum of three different primes. Two-fifths of the pupils gave 15 as the answer. This is the sum of three different primes, but fails to be the correct answer as it is not itself a prime number.

Question 12 does not need algebra. In fact, if you check the options in turn, you should notice very quickly that $2\frac{1}{3} \neq 2 + \frac{1}{3} \times 2\frac{1}{3}$, but $3 = 2 + \frac{1}{3} \times 3$, and so 3 is the correct option. Over half the pupils seem to have just seen '2 kg' and 'a third' in the question and chose $2\frac{1}{3}$ as their answer without further thought.

As usual, it was good to see the number of pupils who achieved very high scores. They deserve our congratulations.

The profile of marks obtained is shown below.

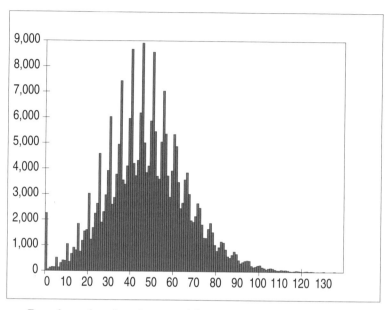

Bar chart showing the actual frequencies in the 2015 JMC

On the basis of the standard proportions used by the UKMT, the cut-off marks were set at

GOLD – 78 or over   SILVER – 62 to 77   BRONZE – 51 to 61

A sample of one of the certificates is shown on the next page.

The Junior Mathematical Olympiad is the follow-up competition to the Challenge. It was decided that candidates who obtained a JMC score of 105 or over were eligible to take part in the JMO. This resulted in 1041 candidates being invited. In line with other follow-on events, schools were allowed to enter 'unqualified' candidates on payment of an appropriate fee. The number who were entered by this route was 109.

An extra follow-on event was introduced this year. Named the Junior Kangaroo, its format is the same as the Senior Kangaroo. The qualification mark was 88.

22

UK JUNIOR MATHEMATICAL CHALLENGE
2015

of

received a

BRONZE CERTIFICATE

Institute
and Faculty
of Actuaries

Professor Dame Frances Kirwan
Chairman, United Kingdom Mathematics Trust

THE UNITED KINGDOM JUNIOR MATHEMATICAL CHALLENGE

The Junior Mathematical Challenge (JMC) is run by the UK Mathematics Trust. The JMC encourages mathematical reasoning, precision of thought, and fluency in using basic mathematical techniques to solve interesting problems. It is aimed at pupils in years 7 and 8 in England and Wales, S1 and S2 in Scotland and years 8 and 9 in Northern Ireland. The problems on the JMC are designed to make students think. Most are accessible, yet challenge those with more experience; they are also meant to be memorable and enjoyable.

Mathematics controls more aspects of the modern world than most people realise – from iPods, cash machines, telecommunications and airline booking systems to production processes in engineering, efficient distribution and stock-holding, investment strategies and 'whispering' jet engines. The scientific and industrial revolutions flowed from the realisation that mathematics was both the language of nature, and also a way of analysing – and hence controlling – our environment. In the last fifty years, old and new applications of mathematical ideas have transformed the way we live.

All of these developments depend on mathematical thinking – a mode of thought whose essential style is far more permanent than the wave of technological change which it has made possible. The problems on the JMC reflect this style, which pervades all mathematics, by encouraging students to think clearly about challenging problems.

The UK JMC has grown out of a national challenge first run in 1988. In recent years over 250,000 pupils have taken part from around 3,700 schools. Certificates are awarded to the highest scoring 40% of candidates (Gold : Silver : Bronze 1 : 2 : 3). From 2014, Certificates of Participation were awarded to all participants.

There is an Intermediate and Senior version for older pupils. All three events are organised by the United Kingdom Mathematics Trust and are administered from the School of Mathematics at the University of Leeds.

The UKMT is a registered charity. For more information about us please visit our website at www.ukmt.org.uk

Donations to support our work would be gratefully received and can be made at www.donate.ukmt.org.uk

The Junior Mathematical Olympiad

UK Junior Mathematical Olympiad 2015

Organised by The United Kingdom Mathematics Trust

Tuesday 9th June 2015

RULES AND GUIDELINES :
READ THESE INSTRUCTIONS CAREFULLY BEFORE STARTING

1. Time allowed: 2 hours.

2. **The use of calculators, measuring instruments and squared paper is forbidden.**

3. All candidates must be in *School Year 8 or below* (England and Wales), *S2 or below* (Scotland), *School Year 9 or below* (Northern Ireland).

4. For questions in Section A *only the answer is required*. Enter each answer neatly in the relevant box on the Front Sheet. Do not hand in rough work. Write in blue or black pen or pencil.

    For questions in Section B you must give *full written solutions*, including clear mathematical explanations as to why your method is correct.

    Solutions must be written neatly on A4 paper. Sheets must be STAPLED together in the top left corner with the Front Sheet on top.

    Do not hand in rough work.

5. Questions A1-A10 are relatively short questions. Try to complete Section A within the first 30 minutes so as to allow well over an hour for Section B.

6. Questions B1-B6 are longer questions requiring *full written solutions*.
    This means that each answer must be accompanied by clear explanations and proofs.
    Work in rough first, then set out your final solution with clear explanations of each step.

7. These problems are meant to be challenging! Do not hurry. Try the earlier questions in each section first (they tend to be easier). Try to finish whole questions even if you are not able to do many. A good candidate will have done most of Section A and given solutions to at least two questions in Section B.

8. Answers must be FULLY SIMPLIFIED, and EXACT using symbols like π, fractions, or square roots if appropriate, but NOT decimal approximations.

DO NOT OPEN THE PAPER UNTIL INSTRUCTED BY THE INVIGILATOR TO DO SO!

The United Kingdom Mathematics Trust is a Registered Charity.

Section A

Try to complete Section A within 30 minutes or so. Only answers are required.

A1. It is 225 minutes until midnight. What time is it on a 24-hour digital clock?

A2. The diagram shows what I see when I look straight down on the top face of a non-standard cubical die. A positive integer is written on each face of the die. The numbers on every pair of opposite faces add up to 10. What is the sum of the numbers on the faces I cannot see?

A3. The diagram shows one square inside another. The perimeter of the shaded region has length 24 cm.
What is the area of the larger square?

A4. My fruit basket contains apples and oranges. The ratio of apples to oranges in the basket is 3 : 8. When I remove one apple the ratio changes to 1 : 3.
How many oranges are in the basket?

A5. Two circles of radius 1 cm fit exactly between two parallel lines, as shown in the diagram. The centres of the circles are 3 cm apart.
What is the area of the shaded region bounded by the circles and the lines?

A6. There are 81 players taking part in a knock-out quiz tournament. Each match in the tournament involves 3 players and only the winner of the match remains in the tournament – the other two players are knocked out. How many matches are required until there is an overall winner?

A7. The diagram shows an equilateral triangle inside a regular hexagon that has sides of length 14 cm. The vertices of the triangle are midpoints of sides of the hexagon.
What is the length of the perimeter of the triangle?

A8. What is the units digit in the answer to the sum $9^{2015} + 9^{2016}$?

A9. The figure shows part of a tiling, which extends indefinitely in every direction across the whole plane. Each tile is a regular hexagon. Some of the tiles are white, the others are black.
What fraction of the plane is black?

A10. Lucy wants to put the numbers 2, 3, 4, 5, 6 and 10 into the circles so that the products of the three numbers along each edge are the same, and as large as possible.
What is this product?

Section B

Your solutions to Section B will have a major effect on your JMO result. Concentrate on one or two questions first and then **write out full solutions** (not just brief 'answers').

B1. Let N be the smallest positive integer whose digits add up to 2015.
What is the sum of the digits of $N + 1$?

B2. The diagram shows triangle ABC, in which $\angle ABC = 72°$ and $\angle CAB = 84°$. The point E lies on AB so that EC bisects $\angle BCA$. The point F lies on CA extended. The point D lies on CB extended so that DA bisects $\angle BAF$.

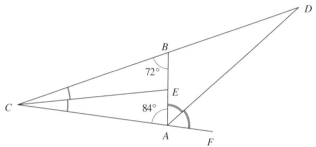

Prove that $AD = CE$.

B3. Jack starts in the small square shown shaded on the grid, and makes a sequence of moves. Each move is to a neighbouring small square, where two small squares are neighbouring if they have an edge in common. He may visit a square more than once.

Jack makes four moves. In how many different small squares could Jack finish?

B4. The point F lies inside the regular pentagon $ABCDE$ so that $ABFE$ is a rhombus. Prove that EFC is a straight line.

B5. I have two types of square tile. One type has a side length of 1 cm and the other has a side length of 2 cm.
What is the smallest square that can be made with equal numbers of each type of tile?

B6. The letters a, b, c, d, e and f represent single digits and each letter represents a different digit. They satisfy the following equations:

$$a + b = d, \qquad b + c = e \qquad \text{and} \qquad d + e = f.$$

Find all possible solutions for the values of a, b, c, d, e and f.

UK Junior Mathematical Olympiad 2015 Solutions

A1 20:15 The 225 minutes are equal to 3 hours and 45 minutes. At 3 hours before midnight the time is 21:00. So 45 minutes earlier than that the time is 20:15.

A2 23 A cubical die has six faces and so three pairs of opposite faces. The numbers on each pair of opposite faces add to 10, so the sum of the numbers on all six faces is 30. Therefore the sum of the numbers on the unseen faces is $30 - 7 = 23$.

A3 36 cm² The length of the perimeter of the shaded region is the same as the length of the perimeter of the square. Hence the square has side length 6 cm and so has area 36 cm².

A4 24 Suppose there are initially m apples and n oranges in the basket. Since $m : n = 3 : 8$, it follows that $8m = 3n$. After removing one apple, the ratio becomes $(m - 1) : n = 1 : 3$, and so $3(m - 1) = n$. Putting these two equations together,

$$8m = 3 \times 3(m - 1)$$
$$8m = 9m - 9$$
$$m = 9.$$

Using the second equation gives $n = 3 \times 8 = 24$. So there are 24 oranges in the basket.

A5 $(6 - \pi)$ cm² The area bounded by the circles can be calculated by subtracting the area of one circle from the area of the 3 cm × 2 cm rectangle with edges passing through the centres of the circle. Therefore the shaded area is $(6 - \pi)$ cm².

A6 40 Two players are knocked out as a result of one match. To leave a winner, 80 players must be knocked out. Therefore there must be 40 matches.

A7 63 cm The hexagon may be divided into small equilateral triangles as shown.

Each of these small triangles has side length 7 cm and so the perimeter of the shaded triangle is 9×7 cm $= 63$ cm.

A8 0 The sum can be factorised to give $9^{2015} + 9^{2016} = 9^{2015}(1 + 9)$. This is equal to $9^{2015} \times 10$ and so, since this number is a multiple of 10, its units digit must be 0.

A9 $\frac{1}{8}$ The tiling pattern can be divided into identical pieces, each of which has the following shape.

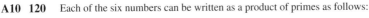

As $\frac{1}{8}$ of each piece is black, it follows that $\frac{1}{8}$ of the plane is black.

A10   **120**   Each of the six numbers can be written as a product of primes as follows:

$$2, \qquad 3, \qquad 2 \times 2, \qquad 5, \qquad 2 \times 3, \qquad 2 \times 5.$$

Since only two 3s appear in the list, the numbers 3 and 6 cannot appear in the same edge of the triangle; otherwise the products of the three numbers along this edge would have a factor of 3^2 and product of the three numbers along the other two edges would not. The same is true for 5 and 10 since there are only two 5s in the list. These observations mean that, up to symmetry, the only possible arrangements are:

For each of these arrangements, the 2 and the 4 can be placed in two ways to give the following eight arrangements.

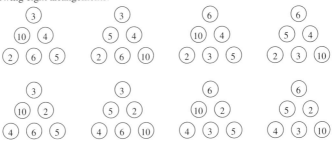

Consider the products of the numbers along the edges of each of these arrangements. In the first arrangement, the products are all 60; in the last, the products are all 120; in each of the others, the products are not all equal. Therefore the largest possible product satisfying the conditions of the question is 120.

28

B1 Let N be the smallest positive integer whose digits add up to 2015.
What is the sum of the digits of $N + 1$?

Solution

For such an integer to be as small as possible, it must have as few digits as possible (since any integer with more digits would be larger).

Since $\frac{2015}{9} = 223$ remainder 8, the smallest possible number of digits is 224. Thus any number whose digits are made up of 223 copies of '9' and one '8' will have the correct digit sum and use the smallest possible number of digits. The integer N must have '8' as its leading digit, followed by 223 copies of '9'; any other arrangement of these digits would give a larger integer.

Thus $N + 1 = 9 \times 10^{223}$. Hence the digit sum of $N + 1$ is 9.

B2 The diagram shows triangle ABC, in which $\angle ABC = 72°$ and $\angle CAB = 84°$. The point E lies on AB so that EC bisects $\angle BCA$. The point F lies on CA extended. The point D lies on CB extended so that DA bisects $\angle BAF$.

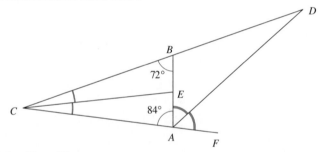

Prove that $AD = CE$.

Solution

Look first at triangle ABC. Since the angles in a triangle add to 180°, $\angle BCA = 24°$. Hence, since CE bisects $\angle BCA$, $\angle BCE = \angle ECA = 12°$. Next consider triangle ECA and use the angle sum again to obtain $\angle AEC = 84°$. Therefore $\angle CAE = \angle AEC$ and so triangle ECA is isosceles and hence $AC = CE$.

Consider the angles at A. Since angles on a straight line add to 180°, $\angle BAF = 96°$. Because AD bisects $\angle BAF$, $\angle BAD = 48°$. Finally, consider triangle DCA. Since the angles in a triangle add to 180°, it must be the case that $\angle ADC = 24°$. Therefore $\angle BCA = \angle ADC$ and so triangle DCA is isosceles and hence $AC = AD$. Therefore $AD = CE$ since both are equal to AC.

B3   Jack starts in the small square shown shaded on the
grid, and makes a sequence of moves. Each move is to
a neighbouring small square, where two small squares
are neighbouring if they have an edge in common. He
may visit a square more than once.

Jack makes four moves. In how many different small
squares could Jack finish?

Solution

Let the starting square be labelled (0, 0) and the square to the right of the starting square be
labelled (1, 0), and so on for the remaining squares in the grid.

Each move may be represented by (+1, +0), (+0, +1), (−1, +0) or (+0, −1) where, for
example, (+1, +0) represents a move of one square to the right and (+0, −1) represents a
move of one square downwards.

Each move increases or decreases the sum of the coordinates of the occupied square by 1. At
(0, 0) the sum of the coordinates is 0. After one move, the sum of the coordinates must be 1
or −1. So after two moves the sum of the coordinates is 2, 0 or −2. After three moves, the
sum of the coordinates of the occupied square is 3, 1, −1 or −3. Finally, after the fourth move,
the sum of the coordinates of the occupied square is 4, 2, 0, −2, or −4.

This means the sequence of four moves can end at any one of the 25 squares shown in black.

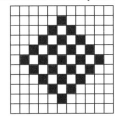

B4 The point F lies inside the regular pentagon $ABCDE$ so that $ABFE$ is a rhombus. Prove that EFC is a straight line.

Solution

Each interior angle of a regular pentagon is $108°$. The internal angles of a quadrilateral sum to $360°$ and so, since $ABFE$ is a rhombus, $\angle ABF = \angle FEA = 72°$. Therefore $\angle FBC = 36°$. Triangle FBC is isosceles since $BC = AB = BF$ and so $\angle BFC = \angle BCF = 72°$. Then $\angle EFC = \angle EFB + \angle BFC = 108° + 72° = 180°$ and so EFC is a straight line.

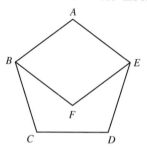

B5 I have two types of square tile. One type has a side length of 1 cm and the other has a side length of 2 cm.

What is the smallest square that can be made with equal numbers of each type of tile?

Solution

Each of the 1 cm tiles has area 1 cm^2 and each of the 2 cm tiles has area 4 cm^2. Suppose there are n tiles of each type. Then the assembled square must have area $5n$ cm^2 and $5n$ must be a square number.

The smallest n could be is 5. This would require the tiles to fit together to form a 5 cm × 5 cm square. However the first four 2 cm tiles are placed in a 5 cm square, the fifth 2 cm tile cannot be placed in the remaining space. So n cannot be 5.

The next smallest n for which $5n$ is a square is 20. This would require the tiles to fit together to form a 10 cm × 10 cm square. This arrangement is possible; an example is shown below. Hence the smallest square that can be made is a 10 cm × 10 cm square.

B6   The letters a, b, c, d, e and f represent single digits and each letter represents a different digit. They satisfy the following equations:

$$a + b = d, \qquad b + c = e \qquad \text{and} \qquad d + e = f.$$

Find all possible solutions for the values of a, b, c, d, e and f.

Solution

The equations may be represented as the following triangle where each digit is the sum of the two adjacent digits directly above it.

$$\begin{array}{ccc} a & b & c \\ & d \quad e & \\ & f & \end{array}$$

None of the digits a, b, c, d and e may be 0 since this would force two of the others to be equal. The digit f cannot be 0 since $f = d + e$ and both d and e are positive.

Consider the middle row and note that both d and e must be at least 3 since they are each the sum of distinct positive integers. Now assume that $d < e$. Since $d \geqslant 3$ and $f \leqslant 9$ it follows that $e \leqslant 6$. This means that (d, e) is $(3, 4)$, $(3, 5)$, $(3, 6)$ or $(4, 5)$. However, the only way to write 4 as the sum of two different nonzero digits is $1 + 3$; therefore $(d, e) \neq (3, 4)$.

Suppose $(d, e) = (3, 5)$.

$$\begin{array}{ccc} a & b & c \\ & 3 \quad 5 & \\ & 8 & \end{array}$$

Then $b \neq 2$ since this would mean $c = 3$. So it must be the case that $(a, b, c) = (2, 1, 4)$.

Now suppose $(d, e) = (3, 6)$.

$$\begin{array}{ccc} a & b & c \\ & 3 \quad 6 & \\ & 9 & \end{array}$$

Then a must be 1 or 2 and so (a, b, c) is $(1, 2, 4)$ or $(2, 1, 5)$.

Finally, suppose $(d, e) = (4, 5)$.

$$\begin{array}{ccc} a & b & c \\ & 4 \quad 5 & \\ & 9 & \end{array}$$

Then a must be 1 or 3. But a cannot be equal to 3 as this would force c to be 4. So $(a, b, c) = (1, 3, 2)$.

Therefore, with the assumption that $d < e$, there are four solutions. Another four can be obtained by assuming $d > e$, which has the effect of reflecting each of the triangles vertically.

The solutions for (a, b, c, d, e, f) are

$(2, 1, 4, 3, 5, 8)$,	$(1, 2, 4, 3, 6, 9)$,	$(2, 1, 5, 3, 6, 9)$,	$(1, 3, 2, 4, 5, 9)$,
$(4, 1, 2, 5, 3, 8)$,	$(4, 2, 1, 6, 3, 9)$,	$(5, 1, 2, 6, 3, 9)$,	$(2, 3, 1, 5, 4, 9)$.

The marking and results

The pupils' scripts began arriving very rapidly and the marking took place in Leeds on the weekend of 20th and 21st June. The discussions as to how marks should be given and for what were ably led by Steven O'Hagan. A full list of markers appears in the Volunteers section.

As has been stated, the object of the JMO is for pupils to be *challenged*, possibly in ways they have not been before. Some participants may find all of Section B rather beyond them, but it is hoped that they achieve a degree of satisfaction from Section A. Satisfaction is an important aspect of this level of paper; nevertheless, those who do succeed in tackling Section B deserve credit for that and such credit is mainly dependent on getting solutions to questions in Section B which are 'perfect' or very nearly so.

Based on the total of both A and B sections, book prizes and certificates are awarded. The top scoring 25% of all candidates receive a Certificate of Distinction. Of those scoring below this, candidates who make a good attempt at the paper (usually a score of 8+, but may be moderated based on the score distribution) will receive a Certificate of Merit. Of the remaining candidates, those who qualified automatically for the JMO via the JMC receive a Certificate of Qualification. Other candidates (discretionary) are only eligible for Distinction and Merit certificates. The top 50 scorers will be awarded a book prize; in 2015 this was *The Number Devil* by Hans Enzenberger.

Medals are awarded on a different basis to certificates and prizes. The medal-awarding total will be the Section A mark + Section B marks that are 5 or more. The top 210 will receive a medal; gold: silver: bronze 30:60:120.

Average marks were similar to 2014. The numbers of medals awarded were: 29 Gold, 57 Silver and 130 Bronze.

The list below includes all the medal winners in the 2015 JMO. Within each category, the names are in alphabetical order.

Special mention should be made of Freddie Hand, Tommy Mackay and Alex Yan each of whom now have two JMO gold medals.

The results and all the extras (books, book plates, certificates and medals) were posted to schools by the middle of July. Where appropriate, some materials were e-mailed to schools.

GOLD MEDALS

Charles Anderton	St Olave's Grammar School, Kent
Wilfred Ashworth	Sutton Grammar School for Boys, Surrey
Victor Baycroft	Notre Dame High School, Sheffield
Robin Bradfield	Cargilfield Preparatory, Edinburgh
Olivia Brown	Backwell School, North Somerset
Nathan Burn	Bishop Wordsworth's School, Salisbury
James Chen	Chetham's School of Music, Manchester
Soren Choi	Westminster Under School
David Cowen	Ermysted's Grammar School, N. Yorks
Ben Fearnhead	Lancaster Royal Grammar School
Alex Ford	Ryecroft Middle School, Staffs
Thomas Frith	Horsforth School, Leeds
Freddie Hand	Judd School, Tonbridge, Kent
Andre Heycock	Wychwood School, Oxford
Michael Hua	Beijing Dulwich International School
Otoharu Kawaguchi	Clifton Lodge School, London
Rubaiyat Khondaker	Wilson's School, Surrey
Linus Luu	St Olave's Grammar School, Kent
Fraser Mason	St Mary's Music School, Edinburgh
Dillon Patel	King Edward VI School, Stratford-upon-Avon
Zoe Price	Dollar Academy, Clackmannanshire
Anthony Shin	Caldicott School, Buckinghamshire
Vikram Singh	The Perse School, Cambridge
James Tan	Queen Elizabeth's School, Barnet
Thien Udomsrirungruang	Shrewsbury International School, Thailand
Stephon Umashangar	Hampton School, Middlesex
Tommy Walker Mackay	Stretford Grammar School, Manchester
Alex Yan	Bancroft's School, Essex
Lauren Zhang	King Edward VI H. Sch. for Girls, Birmingham

SILVER MEDALS

Naomi Bazlov	King Edward VI H. Sch. for Girls, Birmingham
Charlie Buckingham	Danes Hill School, Surrey
Matthew Buckley	The Hall School, Hampstead
Dewi Chappel	Hockerill Anglo-European Coll., Bishop's Stortford
Claudia Cilleruelo	City of London Girls' School
Alex Colesmith	Judd School, Tonbridge, Kent
Matthew Cresswell	Hampton School, Middlesex
Brian Davies	St Edward's College, Liverpool
Andy Deng	Wilson's School, Surrey
Rishit Dhoot	West Hill School, Lancashire
Andrew Dubois	Wellsway School, Bristol
Harry Elliot	Highgate School, London
Gil Elstein	University College School, London
Rebekah Fearnhead	Lancaster Girls' Grammar School
Rujun Feng	Beaumont School, St Albans
Rehan Gamage	Wilson's School, Surrey
Anna Griffiths	Howard of Effingham School, Surrey
Maxi Grindley	Hampton School, Middlesex
George Guest	St Olave's Grammar School, Kent
Nathan Halliday	Sawston Village College, Cambridgeshire
Sean Hargraves	Kellett School, Hong Kong
Sarah Henderson	Highgate School, London
Rikako Hirai	St Paul's Girls' School, Hammersmith
Edward Hu	Queen Elizabeth's School, Barnet
Jennifer Hu	Loughborough High School
Matthew Johnson	Kesgrave High School, Ipswich
Ann Kanchanasakdichai	Bangkok Patana School
Fredric Kong	Beijing Dulwich International School
Shai Kuganesan	Queen Elizabeth's School, Barnet
Thomas Kweon	Beijing Dulwich International School
Joe Lee	Drayton Manor High School, London
Alice Li	Old Palace School, Croydon

Tony Lin	Wilson's School, Surrey
Nathan Lockwood	The Sele School, Hertford
Saomiyan Mathetharan	Tiffin School, Kingston-upon-Thames
Vishvesh Mehta	The Portsmouth Grammar School
Ilya Misyura	Westminster Under School
Emre Mutlu	British School of Chicago, Illinois, USA
Euan Ong	Magdalen College School, Oxford
Ollie Perree	Colyton Grammar School, Devon
Matthew Perry	St Olave's Grammar School, Kent
Geno Racklin Asher	University College School, London
Kiran Raja	Manchester Grammar School
Benedict Randall Shaw	Westminster Under School
Peter Rose	Wymondham High School, Norfolk
William Rous	Swavesey Village College, Cambridgeshire
Razik Sheikh	City of London School
Lauren Shlomovich	The Stephen Perse Foundation, Cambridge
Oliver Sier	City of London School
Gemma Taylor	Oxford High School
Aron Thomas	Dame Alice Owen's School, Herts
Imogen Thomas	Highgate School, London
Arseny Uskov	Northwood Prep School, Herts
Harry Vaughan	Tiffin School, Kingston-upon-Thames
Eric Yin	Dulwich College Shanghai
Yoshie Yuki	Clifton Lodge School, London
Anna Zhou	Royal High School, Bath

BRONZE MEDALS

Robert Ackroyd	Lancaster Royal Grammar School
Alimzhan Adil	Regent International School Dubai
Suyash Agarwal	St John's College, Cardiff
Yuvraj Agarwal	Sutton Grammar School for Boys, Surrey
Mohit Agarwalla	Tiffin School, Kingston-upon-Thames
Ayham Alkhader	Colet Court School, London

Edward Allen	Hampton School, Middlesex
Gaurav Arya	King George V School, Hong Kong
Arthur Ashworth	Sutton Grammar School for Boys, Surrey
Isaac Backhouse	Silcoates School, Wakefield
Ensol Baek	North London Collegiate Sch. Jeju, South Korea
Ravi Bahukhandi	British School Manila, Phillipines
Gabriel Ball	Highgate School, London
Neha Banerjee	Fullbrook School, Surrey
Venusha Baskarathasan	St Helen's School, Middlesex
Agathiayan Bragadeesh	Hymers College, Hull
Mark Bramble	King Edward VI Five Ways Sch., Birmingham
Jude Burling	The Perse School, Cambridge
John Carlyon	City of London School
Christian Cases	Tiffin School, Kingston-upon-Thames
Sunay Challa	Queen Elizabeth's School, Barnet
Renee Chang	King Edward VI HS for Girls, Birmingham
Xue Bang Chen	King Edward VI Camp Hill Boys' S, Birmingham
Anthony Chen	Sha Tin College, Hong Kong
William Ching	The Hall School, Hampstead
Nicole Choong	Garden International School, Malaysia
Soumyaditya Choudhuri	UWCSEA Dover Campus, Singapore
Sacha Chowdhury	Highgate School, London
James Christy	St Aidan's CE High School, Harrogate
Miranda Connolley	The Perse School, Cambridge
Alex Curran	The Royal Grammar School, High Wycombe
Sam Curtis	Huntington School, York
Sara D'Attanasio	St Paul's Girls' School, Hammersmith
Kira Dhariwal	Wycombe High School, Buckinghamshire
William Downes	Birkdale School, Sheffield
Alec Durgheu	Latymer Upper School, Hammersmith
James Edmiston	Magdalen College School, Oxford
Joel Fair	Bethany School, Sheffield
Anton Fedotov	Colet Court School, London
Cameron Fraser	Merchant Taylors' School, Middlesex

Richard Gong	The Romsey School, Hampshire
Anant Gupta	Whitgift School, Surrey
Abbas Habib Kazmi	Ilford County High School, Essex
Shenuka Haegoda	Pate's Grammar School, Cheltenham
Nathan Hardcastle	St Joseph's RC Middle S., Northumberland
Joel Harris	University College School, London
Will Haskins	Wells Cathedral School, Somerset
Ben Howell	Berkhamsted Collegiate S. (Castle), Herts
Tian Hsu	Bute House Prep School
John Hu	West Island School (ESF), Hong Kong
Kevin Huang	Dulwich College Suzhou, China
Jeffery Huang	Dulwich College Suzhou, China
Maximo Ivakhno	International College San Pedro, Spain
Freddy Jiang	Renaissance College, Hong Kong
Elana Jobson	Swanmore Technology College, Hampshire
Sam Ketchell	Weaverham High School, Cheshire
Hayyan Khan	Watford Grammar School for Boys
Subin Kim	North London Collegiate S Jeju, South Korea
Yoo Hyun Kim	The Regent's School Pattaya, Thailand
Yoon Jin Kim	St Paul's Girls' School, Hammersmith
Daniel Knight	Imberhorne School, East Grinstead
Sebastian K. Wellsted	Dame Alice Owen's School, Herts
Avish Kumar	Westminster Under School
Ryan Kwak	North London Collegiate S. Jeju, South Korea
Kai Lam	Whitgift School, Surrey
Seojoon Lee	Overseas Family School, Singapore
Sangmin Lee	North London Collegiate S. Jeju, South Korea
Austen Leitch	Blenheim High School, Epsom
Sophie Leman	Beaconsfield High School, Buckinghamshire
James Lester	Judd School, Tonbridge, Kent
Jiajiong Liu	Manchester Grammar School
Jack Lucas	Hampton School, Middlesex
Vanika Mahesh	UWCSEA Dover Campus, Singapore
Shivendu Mandal	King Edward's School, Birmingham

Neel Maniar	Wilson's School, Surrey
Callum Martin	Comberton Village College, Cambridgeshire
David Maxen	Parmiter's School, Watford
Ruairi McCabe	Bolton School (Boys Division)
Maxwell Mckeay	Fullbrook School, Surrey
Lydia Mekonnen	Henrietta Barnett School, London
Harry Melling	Queen Elizabeth High School, Gainsborough
Kai Mitsuishi	Bangkok Patana School
Seb Mobus	Tapton School, Sheffield
Isaac Mort	Lancaster Royal Grammar School
Jack Murphy	Hampton School, Middlesex
Ewan Neumann	Alleyn's School, Dulwich
Daniel Newton	De Aston School, Lincolnshire
Max Northcott	Roundwood Park School, Harpenden
Chaewon Oh	British International S. Ho Chi Minh City
Robert Peacock	Queen Mary's Grammar School, Walsall
Jack Peck	Soham Village College, Cambridgeshire
Vlad Penzyev	Hampton School, Middlesex
Tristan Peroy	City of London School
William Pizii	Hereford Cathedral School
Xander Povey	Torquay Boys' Grammar School
Chenxin Qiu	Douglas Academy, East Dunbartonshire
Mukund Raghavan	King Edward VI School, Southampton
Diogo Ramos	Ibstock Place School, Roehampton, Surrey
Luke Remus Elliot	Westminster Under School
Luke Rennells	Simon Langton Boys' Grammar S., Canterbury
Jessica Richards	South Wilts Grammar School, Salisbury
Joanne Roper	Park Community School, Barnstaple
Tibor Rothschild	University College School, London
James Saker	Jewish Community Secondary School, Barnet
Oscar Selby	Westminster Under School
Drew Sellis	Queen Elizabeth's School, Barnet
Shrey Shah	Northwood Prep School, Herts
Luke Sharkey	St Andrew's and St Bride's HS, East Kilbride

Miles Shelley	King Edward's School, Birmingham
Matilda Sidel	St Paul's Girls' School, Hammersmith
Aryan Singh	George Heriot's School, Edinburgh
Grace Slattery	English Martyrs Catholic Voluntary A, Rutland
Andrew Spielmann	Colet Court School, London
Paris Suksmith	Harrow International School, Bangkok
Jeffrey Tan	Aylesbury Grammar School
Ashwin Tennant	Abingdon Prep School, Oxon
Luxmi Thayaparan	North London Collegiate School
Louis Thomson	Lanesborough School, Guildford
Amu Varma	Colet Court School, London
Satyam Verma	Birkdale School, Sheffield
Luis Wahl	Colet Court School, London
Alex Walker	The Perse School, Cambridge
Lucy Wang	Watford Grammar School for Girls
Isaac Weaver	Ipswich School
David Xu	Merchant Taylors' Boys School, Liverpool
Ray Yan	Latymer School, London
Andrew Yang	Reading School
Jenny Yang	Bournemouth School for Girls
Hanyu Yin	Wolverhampton Girls' High School
Alex Zeier	Colet Court School, London

The next stage

In 2015, we introduced a Kangaroo follow-on round to the Junior Maths Challenge, to run on the same day as the Junior Olympiad. We now have both an Olympiad and Kangaroo follow-on for all of our Challenges. The Junior Kangaroo is a one-hour multiple-choice paper with 25 questions for the UK and by invitation only. It was offered to around 5,000 UK candidates who scored just below the Junior Olympiad qualifying score. The qualification mark was 88.

Junior Kangaroo Mathematical Challenge

Tuesday 9th June 2015

Organised by the United Kingdom Mathematics Trust

The Junior Kangaroo allows students in the UK to test themselves on questions set for young mathematicians from across Europe and beyond.

RULES AND GUIDELINES (to be read before starting):

1. Do not open the paper until the Invigilator tells you to do so.

2. Time allowed: **1 hour**.
   No answers, or personal details, may be entered after the allowed hour is over.

3. The use of rough paper is allowed; **calculators** and measuring instruments are **forbidden**.

4. Candidates in England and Wales must be in School Year 8 or below.
   Candidates in Scotland must be in S2 or below.
   Candidates in Northern Ireland must be in School Year 9 or below.

5. **Use B or HB pencil only**. For each question mark *at most one* of the options A, B, C, D, E on the Answer Sheet. Do not mark more than one option.

6. Five marks will be awarded for each correct answer to Questions 1 - 15.
   Six marks will be awarded for each correct answer to Questions 16 - 25.

7. *Do not expect to finish the whole paper in 1 hour*. Concentrate first on Questions 1-15. When you have checked your answers to these, have a go at some of the later questions.

8. The questions on this paper challenge you **to think**, not to guess. Though you will not lose marks for getting answers wrong, you will undoubtedly get more marks, and more satisfaction, by doing a few questions carefully than by guessing lots of answers.

Enquiries about the Junior Kangaroo should be sent to: Maths Challenges Office,
School of Mathematics, University of Leeds, Leeds, LS2 9JT.
(Tel. 0113 343 2339)
http://www.ukmt.org.uk

1. Ben lives in a large house with his father, mother, sister and brother as well as 2 dogs, 3 cats, 4 parrots and 5 goldfish. How many legs are there in the house?

   A 18   B 36   C 38   D 46   E 66

2. The sum of five consecutive integers is 2015. What is the smallest of these integers?

   A 401   B 403   C 405   D 407   E 409

3. The diagram on the right shows a cube of side 18 cm. A giant ant walks across the cube's surface from X to Y along the route shown. How far does it walk?

   A 54 cm   B 72 cm   C 80 cm   D 88 cm   E 90 cm

4. How many seconds are there in $\frac{1}{4}$ of $\frac{1}{6}$ of $\frac{1}{8}$ of a day?

   A 60   B 120   C 450   D 900   E 3600

5. What is $203\,515 \div 2015$?

   A 11   B 101   C 1001   D 111   E 103

6. In the diagram, five rectangles of the same size are shown with each side labelled with a number.

A	B	C	D	E
7 5 4 8	3 8 5 0	9 0 7 2	1 2 3 6	4 1 9 6

   These rectangles are placed in the positions I to V as shown so that the numbers on the sides that touch each other are equal.

	I	II	III
	IV	V	

   Which of the rectangles should be placed in position I?

   A   B   C   D   E

7. Selina takes a sheet of paper and cuts it into 10 pieces. She then takes one of these pieces and cuts it into 10 smaller pieces. She then takes another piece and cuts it into 10 smaller pieces and finally cuts one of the smaller pieces into 10 tiny pieces. How many pieces of paper has the original sheet been cut into?

   A 27   B 30   C 37   D 40   E 47

8. John takes 40 minutes to walk to school and then to run home. When he runs both ways, it takes him 24 minutes. He has one fixed speed whenever he walks, and another fixed speed whenever he runs. How long would it take him to walk both ways?

   A 56 minutes   B 50 minutes   C 44 minutes   D 28 minutes   E 24 minutes

9. In the diagram on the right, the number in each circle is the sum of the numbers in the two circles below it. What is the value of *x*?

A 100   B 82   C 55   D 50   E 32

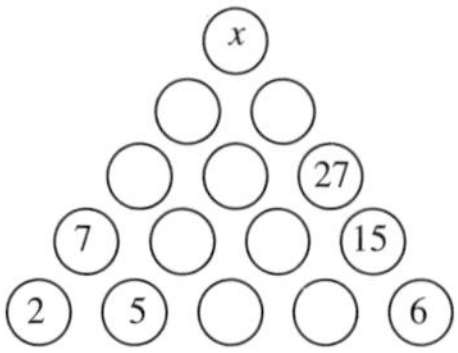

10. The diagram on the right shows a large triangle divided up into squares and triangles. *S* is the number of squares of any size in the diagram and *T* is the number of triangles of any size in the diagram. What is the value of $S \times T$?

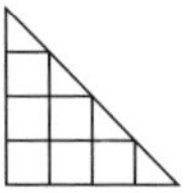

A 30   B 35   C 48   D 70   E 100

11. In the diagram, the small equilateral triangles have area 4 cm². What is the area of the shaded region?

A 80 cm² B 90 cm² C 100 cm² D 110 cm² E 120 cm²

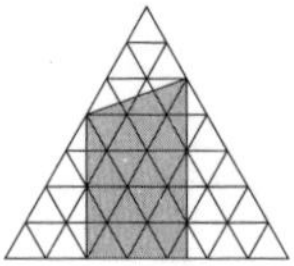

12. In the sum shown, different shapes represent different digits. What digit does the square represent?

A 2   B 4   C 6   D 8   E 9

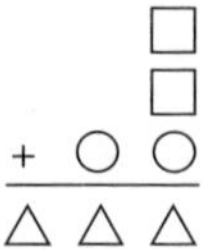

13. The sum of 10 distinct positive integers is 100. What is the largest possible value of any of the 10 integers?

A 55   B 56   C 60   D 65   E 91

14. The diagram shows five circles of the same radius touching each other. A square is drawn so that its vertices are at the centres of the four outer circles.

What is the ratio of the area of the shaded parts of the circles to the area of the unshaded parts of the circles?

A 1:3   B 1:4   C 2:5   D 2:3   E 5:4

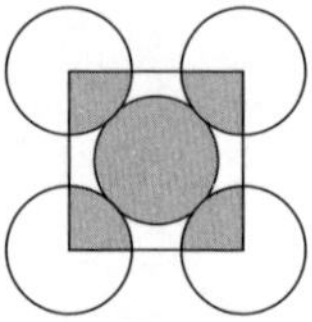

15. A rectangular garden is surrounded by a path of constant width. The perimeter of the garden is 24 m shorter than the distance along the outside edge of the path. What is the width of the path?

A 1 m   B 2 m   C 3 m   D 4 m   E 5 m

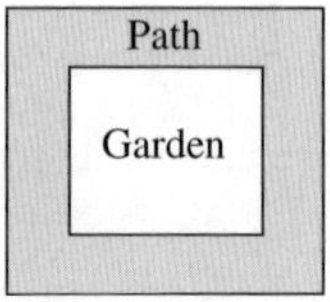

16. A caterpillar starts from its hole and moves across the ground, turning 90° either left or right after each hour. It moves 2 m in the first hour, followed by 3 m in the second hour and 4 m in the third hour and so on. What is the greatest distance it can be from its hole after seven hours?

A 35 m   B 30 m   C 25 m   D 20 m   E 15 m

17. In a pirate's trunk there are 5 chests. In each chest there are 4 boxes and in each box there are 10 gold coins. The trunk, the chests and the boxes are all locked. Blind Pew unlocks 9 locks and takes all the coins in all the boxes he unlocks. What is the smallest number of gold coins he could take?

A 20   B 30   C 40   D 50   E 70

18. Brian chooses an integer, multiplies it by 4 then subtracts 30. He then multiplies his answer by 2 and finally subtracts 10. His answer is a two-digit number. What is the largest integer he could choose?

A 10   B 15   C 18   D 20   E 21

19. From noon till midnight, Clever Cat sleeps under the oak tree and from midnight till noon he is awake telling stories. A poster on the tree above him says "Two hours ago, Clever Cat was doing the same thing as he will be doing in one hour's time". For how many hours a day does the poster tell the truth?

A 3   B 6   C 12   D 18   E 21

20. The diagram below shows a sequence of shapes made up of black and white floor tiles where each shape after the first has two more rows and two more columns than the one before it.

1st   2nd   3rd   15th

How many black tiles would be required to create the 15th shape in the sequence?

A 401   B 421   C 441   D 461   E 481

21. Peter has a lock with a three-digit code. He knows that all the digits of his code are different and that if he divides the second digit by the third and then squares his answer, he will get the first digit. What is the difference between the largest and smallest possible codes?

A 42   B 468   C 499   D 510   E 541

22.

The diagram above shows the front and right-hand views of a solid made up of cubes of side 3 cm. The maximum volume that the solid could have is V cm^3. What is the value of V?

A 162   B 216   C 324   D 540   E 648

23. How many three-digit numbers have an odd number of factors?

A 5   B 10   C 20   D 21   E 22

24. Molly, Dolly, Sally, Elly and Kelly are sitting on a park bench. Molly is not sitting on the far right and Dolly is not sitting on the far left. Sally is not sitting at either end. Kelly is not sitting next to Sally and Sally is not sitting next to Dolly. Elly is sitting to the right of Dolly but not necessarily next to her. Who is sitting at the far right end?

A Molly   B Dolly   C Sally   D Kelly   E Elly

25. Anna, Bridgit and Carol run in a 100 m race. When Anna finishes, Bridgit is 16 m behind her and when Bridgit finishes, Carol is 25 m behind her. The girls run at constant speeds throughout the race. How far behind was Carol when Anna finished?

A 37 m   B 41 m   C 50 m   D 55 m   E 60 m

Further remarks

Solutions were provided.

2015 Junior Kangaroo Solutions

1. **C** Ben, his father, his mother, his sister, his brother and all four of the parrots have two legs each, making 18 legs in total. The two dogs and the three cats have four legs each, making 20 legs in total. Hence there are 38 legs in the house.

2. **A** Let the five consecutive integers be $n - 2$, $n - 1$, n, $n + 1$ and $n + 2$. These have a sum of $5n$. Hence $5n = 2015$ and therefore $n = 403$. Therefore, the smallest integer is $403 - 2 = 401$.

3. **E** From the diagram, the ant walks the equivalent of five edges. Therefore the ant walks 5×18 cm $= 90$ cm.

4. **C** One day contains 24 hours. Hence $\frac{1}{8}$ of a day is three hours and $\frac{1}{6}$ of this is half an hour. Half an hour contains $\left(\frac{1}{2} \times 60 \times 60\right)$ seconds $= 1800$ seconds and $\frac{1}{4}$ of this is 450 seconds. Hence there are 450 seconds in the required fraction of a day.

5. **B** Long division gives

    $$\begin{array}{r} 101 \\ 2015\overline{)203515} \\ \underline{2015} \\ 2015 \\ \underline{2015} \end{array}$$

    Hence $203\,515 \div 2015 = 101\cdot$

6. **C** Look first at the numbers labelling the left- and right-hand sides of the rectangles. It can be seen that only rectangles A, C and E can be arranged in a row of three with their touching sides equal and so they must form the top row of the diagram. The only common value on the right- and left-hand sides of rectangles B and D is 3 and so rectangle D will be placed in position IV. Therefore, the rectangle to be placed in position I needs to have 2 on its lower edge. Hence rectangle C should be placed in position I (with A in position II, E in position III and B in position V).

7. **C** Each time Selina cuts up a piece of paper, she turns one piece into ten smaller pieces and so the number of pieces she has increases by nine. Selina makes four cuts so the total number of pieces she finishes with is $1 + 4 \times 9 = 37$.

8. **A** It takes John 24 minutes to run both ways so it will take him $\frac{1}{2} \times 24$ minutes $= 12$ minutes to run one way. Also, it takes him 40 minutes to walk one way and run the other so walking one way takes him $\left(40 - 12\right)$ minutes $= 28$ minutes. Hence it will take 2×28 minutes $= 56$ minutes to walk both ways.

9. **B** The empty spaces in the diagram can be completed as shown. Hence the value of x is $35 + 47 = 82$.

10. D In the diagram, there are six 1×1 squares and one 2×2 square. There are also four triangles that are half of a 1×1 square, three triangles that are half of a 2×2 square, two triangles that are half of a 3×3 square and one triangle that is half of a 4×4 square. Hence $S = 7$ and $T = 10$ so $S \times T = 70$.

11. B Let b cm be the length of the base and let h cm be the height of the small equilateral triangles. The area of each triangle is 4 cm^2 so $\frac{1}{2} \times b \times h = 4$.

The shaded area is a trapezium with parallel sides of length $4h$ and $5h$ and with distance $\frac{5}{2}b$ between the parallel lines. Using the formula for the area of a trapezium, the shaded area is equal to $\frac{1}{2}(4h + 5h) \times \frac{5}{2}b = \frac{1}{4} \times 45bh$. From above, $bh = 8$ so this area is equal to $\frac{1}{4} \times 45 \times 8$ cm$^2 = 90$ cm^2.

(Alternatively, one could observe that the first four horizontal rows of the shaded region have area equivalent to five of the small equilateral triangles while the fifth layer has area equivalent to half of that or 2.5 equilateral triangles. Hence the shaded region has area equivalent to 22.5 small equilateral triangles and so has area 22.5×4 cm$^2 = 90$ cm^2.)

12. C The maximum value one can obtain by adding a two-digit number and two one-digit numbers is $99 + 9 + 9 = 117$. Hence the triangle must represent 1 and therefore the sum of the three numbers is 111. To obtain this answer to the sum, the circle must represent 9 since $89 + 9 + 9$ is less than 111. Hence the two squares must sum to $111 - 99 = 12$. Therefore the square represents the digit 6.

13. A To have one of the integers in the sum as large as possible, the other nine must be as small as possible. The minimum possible sum of nine distinct positive integers is $1 + 2 + 3 + 4 + 5 + 6 + 7 + 8 + 9 = 45$. Hence the largest possible integer in the sum is $100 - 45 = 55$.

14. D The shaded area is equal to that of 1 circle $+ 4 \times \frac{1}{4}$ circles $= 2$ circles. The area of the unshaded parts of the circles is equal to that of $4 \times \frac{3}{4}$ circles $= 3$ circles. Hence the required ratio is $2 : 3$.

15. C Let the length and width of the garden be a metres and b metres respectively and let the width of the path be x metres. The perimeter of the garden is $2(a + b)$ metres and the perimeter of the larger rectangle formed by the garden and the path is $2(a + 2x + b + 2x)$ metres. Hence the difference between the distance along the outside edge of the path and the perimeter of the garden in metres is $2(a + 2x + b + 2x) - 2(a + b) = 8x$. Therefore $8x = 24$ which has solution $x = 3$. Hence the width of the path is 3 metres.

16. C The caterpillar will be as far away as possible from its hole if, at each turn, it always heads away from the hole. Hence its maximum distance will occur when it has travelled $(2 + 4 + 6 + 8)$ metres $= 20$ metres in one direction and $(3 + 5 + 7)$ metres $= 15$ metres in a perpendicular direction. Using Pythagoras' Theorem, the maximum distance in metres is then $\sqrt{20^2 + 15^2} = \sqrt{625} = 25$.

17. B To obtain the smallest number of gold coins, the least possible number of boxes must be opened. Therefore, Blind Pew must open the trunk and all five chests, leaving only three boxes to be opened. Hence the smallest number of gold coins he could take is $3 \times 10 = 30$.

18. E Let the integer Brian chooses be x. Following the operations in the question, his final result is $2(4x - 30) - 10$. His answer is a two-digit number so $9 < 2(4x - 30) - 10 < 100$. Hence we have $9 < 8x - 70 < 100$ which has solution $9\frac{7}{8} < x < 21\frac{1}{4}$. Therefore the largest integer Brian could choose is 21.

19. D   Consider the times when the poster does not tell the truth. The poster will not tell the truth one hour before Clever Cat changes his activity and will remain untrue until he has been doing his new activity for two hours. Hence the poster does not tell the truth for three hours around each change of activity but tells the truth the rest of the time. Therefore, the poster will tell the truth for $(24 - 2 \times 3)$ hours $= 18$ hours.

20. B   The nth term of the sequence 1, 3, 5, ... is $2n - 1$. Therefore at the start of the solution the total number of tiles in the 15th shape is $(2 \times 15 - 1)^2 = 29^2 = 841$. In each shape, there is one more black tile than white tile. Hence there would be $\frac{1}{2}(841 + 1) = 421$ black tiles in the 15th shape.

21. E   The digits in the code are all different, so the result of dividing the second digit by the third cannot be 1. The first digit is a square number and, since it cannot be 1, is 4 or 9. The largest possible code will start with 9 and have second digit ÷ third digit = 3 and is 962. The smallest possible code will start with 4 and have second digit ÷ third digit = 2 and is 421. Hence the difference between the largest and smallest possible codes is $962 - 421 = 541$.

22. D   Each cube has volume 3^3 cm^3 $= 27$ cm^3. There are four cubes visible in the base layer from both the front and the side so the maximum number of cubes in the base layer is $4 \times 4 = 16$. Similarly, the maximum number of cubes in the second layer is $2 \times 2 = 4$. Hence the maximum number of cubes in the solid is $16 + 4 = 20$ with a corresponding maximum volume of 20×27 cm^3 $= 540$ cm^3.

23. E   It can be shown that a positive integer has an odd number of factors if and only if it is square. The smallest three-digit square number is $10^2 = 100$ and the largest is $31^2 = 961$. Hence there are $31 - 9 = 22$ three-digit numbers which have an odd number of factors.

24. E   The question tells us that Sally is not sitting at either end. This leaves three possible positions for Sally, which we will call positions 2, 3 and 4 from the left-hand end. Were Sally to sit in place 2, neither Dolly nor Kelly could sit in places 1 or 3 as they cannot sit next to Sally and, since Elly must sit to the right of Dolly, there would be three people to fit into places 4 and 5 which is impossible. Similarly, were Sally to sit in place 3, Dolly could not sit in place 2 or 4 and the question also tells us she cannot sit in place 1 so Dolly would have to sit in place 5 making it impossible for Elly to sit to the right of Dolly. However, were Sally to sit in place 4, Dolly could sit in place 2, Kelly in place 1, Molly (who cannot sit in place 5) in place 3 leaving Elly to sit in place 5 at the right-hand end.

25. A   Carol finishes 25 metres behind Bridgit, so she travels 75 metres while Bridgit runs 100 metres. Therefore she runs 3 metres for every 4 metres Bridgit runs. When Anna finishes, Bridgit has run 84 metres, so that at that time Carol has run $\frac{3}{4} \times 84$ metres $= 63$ metres. Hence Carol finishes $(100 - 63)$ metres $= 37$ metres behind Anna.

The Intermediate Mathematical Challenge and its follow-up events

The Intermediate Mathematical Challenge (IMC) was held on Thursday 5th February 2015. Entries numbered 259,480 and over 211,000 pupils took part. There were several different Intermediate Mathematical Olympiad and Kangaroo (IMOK) follow-up competitions and pupils were invited to the one appropriate to their school year and mark in the IMC. Around 500 candidates in each of Years 9, 10 and 11 sat the Olympiad papers (Cayley, Hamilton and Maclaurin respectively) and approximately 2800 more in each year group took a Kangaroo paper. We start with the IMC paper.

UK INTERMEDIATE MATHEMATICAL CHALLENGE

THURSDAY 5TH FEBRUARY 2015

Organised by the **United Kingdom Mathematics Trust**
and supported by

Institute
and Faculty
of Actuaries

RULES AND GUIDELINES (to be read before starting)

1. Do not open the paper until the Invigilator tells you to do so.

2. Time allowed: **1 hour.**
    No answers, or personal details, may be entered after the allowed hour is over.

3. The use of rough paper is allowed; **calculators** and measuring instruments are **forbidden.**

4. Candidates in England and Wales must be in School Year 11 or below.
    Candidates in Scotland must be in S4 or below.
    Candidates in Northern Ireland must be in School Year 12 or below.

5. **Use B or HB pencil only.** Mark *at most one* of the options A, B, C, D, E on the Answer Sheet for each question. Do not mark more than one option.

6. *Do not expect to finish the whole paper in 1 hour.* Concentrate first on Questions 1-15. When you have checked your answers to these, have a go at some of the later questions.

7. Five marks are awarded for each correct answer to Questions 1-15.
    Six marks are awarded for each correct answer to Questions 16-25.
    Each incorrect answer to Questions 16-20 loses 1 mark.
    Each incorrect answer to Questions 21-25 loses 2 marks.

8. Your Answer Sheet will be read only by a *dumb machine*. **Do not write or doodle on the sheet except to mark your chosen options.** The machine 'sees' all black pencil markings even if they are in the wrong places. If you mark the sheet in the wrong place, or leave bits of rubber stuck to the page, the machine will 'see' a mark and interpret this mark in its own way.

9. The questions on this paper challenge you to **think**, not to guess. You get more marks, and more satisfaction, by doing one question carefully than by guessing lots of answers.
    The UK IMC is about solving interesting problems, not about lucky guessing.

The UKMT is a registered charity
http://www.ukmt.org.uk

1. What is the value of $1 - 0.2 + 0.03 - 0.004$?

   A 0.826   B 0.834   C 0.926   D 1.226   E 1.234

2. Last year, Australian Suzy Walsham won the annual women's race up the 1576 steps of the Empire State Building in New York for a record fifth time. Her winning time was 11 minutes 57 seconds. Approximately how many steps did she climb per minute?

   A 13   B 20   C 80   D 100   E 130

3. What is a half of a third, plus a third of a quarter, plus a quarter of a fifth?

   A $\dfrac{1}{1440}$ B $\dfrac{3}{38}$ C $\dfrac{1}{30}$ D $\dfrac{1}{3}$ E $\dfrac{3}{10}$

4. The diagram shows a regular pentagon inside a square.

   What is the value of x ?

   A 48   B 51   C 54   D 60   E 72

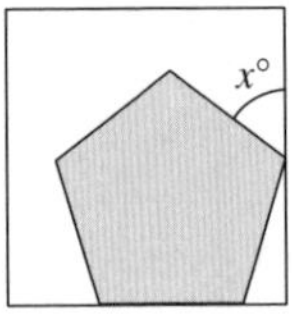

5. Which of the following numbers is not a square?

   A 1^6 B 2^5 C 3^4 D 4^3 E 5^2

6. The equilateral triangle and regular hexagon shown have perimeters of the same length.

   What is the ratio of the area of the triangle to the area of the hexagon?

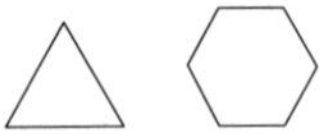

   A $5:6$ B $4:5$ C $3:4$ D $2:3$ E $1:1$

7. A tetrahedron is a solid figure which has four faces, all of which are triangles.

   What is the product of the number of edges and the number of vertices of the tetrahedron?

   A 8   B 10   C 12   D 18   E 24

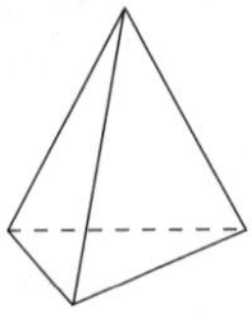

8. How many two-digit squares differ by 1 from a multiple of 10?

   A 1   B 2   C 3   D 4   E 5

9. What is the value of $p + q + r + s + t + u + v + w + x + y$ in the diagram?

   A 540   B 720   C 900   D 1080   E 1440

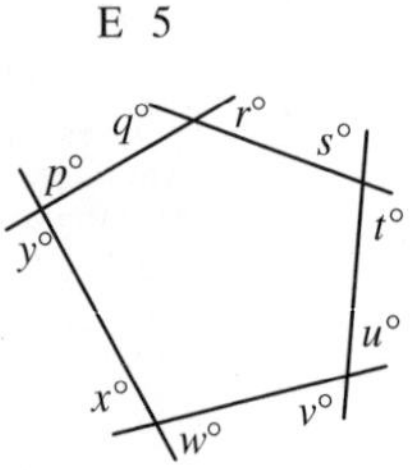

10. What is the remainder when $2^2 \times 3^3 \times 5^5 \times 7^7$ is divided by 8?

    A 2   B 3   C 4   D 5   E 7

11. Three different positive integers have a mean of 7. What is the largest positive integer that could be one of them?

    A 15   B 16   C 17   D 18   E 19

12. An ant is on the square marked with a black dot. The ant moves across an edge from one square to an adjacent square four times and then stops.

    How many of the possible finishing squares are black?

    A 0   B 2   C 4   D 6   E 8

13. What is the area of the shaded region in the rectangle?

    3 cm

    14 cm

    A 21 cm^2 B 22 cm^2 C 23 cm^2 D 24 cm^2 E more information needed

14. In a sequence, each term after the first two terms is the mean of all the terms which come before that term. The first term is 8 and the tenth term is 26. What is the second term?

    A 17   B 18   C 44   D 52   E 68

15. A flag is in the shape of a right-angled triangle, as shown, with the horizontal and vertical sides being of length 72 cm and 24 cm respectively. The flag is divided into 6 vertical stripes of equal width.

    72 cm

    24 cm

    What, in cm^2, is the difference between the areas of any two adjacent stripes?

    A 96   B 72   C 48   D 32   E 24

16. You are asked to choose two positive integers, m and n with $m > n$, so that as many as possible of the expressions $m + n, m - n, m \times n$ and $m \div n$ have values that are prime. When you do this correctly, how many of these four expressions have values that are prime?

    A 0   B 1   C 2   D 3   E 4

17. The football shown is made by sewing together 12 black pentagonal panels and 20 white hexagonal panels. There is a join wherever two panels meet along an edge.

    How many joins are there?

    A 20   B 32   C 60   D 90   E 180

18. The total weight of a box, 20 plates and 30 cups is 4.8 kg. The total weight of the box, 40 plates and 50 cups is 8.4 kg. What is the total weight of the box, 10 plates and 20 cups?

    A 3 kg   B 3.2 kg   C 3.6 kg   D 4 kg   E 4.2 kg

19. The figure shows four smaller squares in the corners of a large
square. The smaller squares have sides of length 1 cm, 2 cm, 3 cm
and 4 cm (in anticlockwise order) and the sides of the large square
have length 11 cm.
What is the area of the shaded quadrilateral?

A 35 cm² B 36 cm² C 37 cm² D 38 cm² E 39 cm²

20. A voucher code is made up of four characters. The first is a letter: V, X or P. The second and
third are different digits. The fourth is the units digit of the sum of the second and third digits.
How many different voucher codes like this are there?

A 180 B 243 C 270 D 300 E 2700

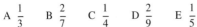

21. A rectangle is placed obliquely on top of an identical rectangle, as
shown.
The area X of the overlapping region (shaded more darkly) is one
eighth of the total shaded area.
What fraction of the area of one rectangle is X ?

A $\frac{1}{3}$ B $\frac{2}{7}$ C $\frac{1}{4}$ D $\frac{2}{9}$ E $\frac{1}{5}$

22. The diagram shows a shaded region inside a large square. The
shaded region is divided into small squares.
What fraction of the area of the large square is shaded?

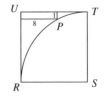

A $\frac{3}{10}$ B $\frac{1}{3}$ C $\frac{3}{8}$ D $\frac{2}{5}$ E $\frac{3}{7}$

23. There are 120 different ways of arranging the letters, U, K, M, I and C. All of these
arrangements are listed in dictionary order, starting with CIKMU. Which position in the list
does UKIMC occupy?

A 110 th B 112 th C 114 th D 116 th E 118 th

24. In square $RSTU$ a quarter-circle arc with centre S is drawn from T
to R. A point P on this arc is 1 unit from TU and 8 units from RU.
What is the length of the side of square $RSTU$?

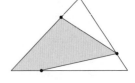

A 9 B 10 C 11 D 12 E 13

25. A point is marked one quarter of the way along
each side of a triangle, as shown.
What fraction of the area of the triangle is
shaded?

A $\frac{7}{16}$ B $\frac{1}{2}$ C $\frac{9}{16}$ D $\frac{5}{8}$ E $\frac{11}{16}$

The IMC solutions

As with the Junior Challenge, a solutions leaflet was sent out.

UK INTERMEDIATE MATHEMATICAL CHALLENGE

THURSDAY 5TH FEBRUARY 2015

Organised by the **United Kingdom Mathematics Trust**
from the School of Mathematics, University of Leeds

SOLUTIONS LEAFLET

This solutions leaflet for the IMC is sent in the hope that it might provide all concerned with some alternative solutions to the ones they have obtained. It is not intended to be definitive. The organisers would be very pleased to receive alternatives created by candidates.

For reasons of space, these solutions are necessarily brief. Extended solutions, and some exercises for further investigation, can be found at:

http://www.ukmt.org.uk/

The UKMT is a registered charity

1. **A** $1 - 0.2 + 0.03 - 0.004 = 0.8 + 0.026 = 0.826.$

2. **E**   The number of steps climbed per minute $\sim \dfrac{1600}{12} = \dfrac{400}{3} \sim 130.$

3. **E**   Half of a third, plus a third of a quarter, plus a quarter of a fifth equals
$$\frac{1}{6} + \frac{1}{12} + \frac{1}{20} = \frac{10 + 5 + 3}{60} = \frac{18}{60} = \frac{3}{10}.$$

4. **C**   The sum of the exterior angles of a convex polygon equals $360°$. The angle marked $p°$ is the exterior angle of a regular pentagon. So $p = 360 \div 5 = 72$. The angle sum of a triangle equals $180°$, so $q = 180 - 90 - 72 = 18$. The angle marked $r°$ is the interior angle of a regular pentagon, so $r = 180 - 72 = 108$. The angles marked $q°$, $r°$ and $x°$ lie along a straight line, so $x = 180 - (q + r)$ $= 180 - (18 + 108) = 54.$

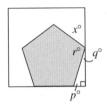

5. **B** $1^6 = (1^3)^2$, $3^4 = (3^2)^2$, $4^3 = 64 = 8^2$ and 5^2 are all squares. However, $2^5 = 32$ and is not a square.

6. **D**   Let the length of the side of the regular hexagon be a. Then its perimeter is $6a$. Therefore the perimeter of the equilateral triangle is also $6a$, so the length of each of its sides is $2a$. The diagrams show that the equilateral triangle may be divided up into 4 equilateral triangles of side a, whereas the regular hexagon may be divided into 6 such triangles. So the required ratio is $4 : 6 = 2 : 3$.

7. **E**   The tetrahedron has 6 edges and 4 vertices, so the required product is $6 \times 4 = 24$.

8. **B**   The two-digit squares are 16, 25, 36, 49, 64 and 81. Of these, only 49 and 81 differ by 1 from a multiple of 10.

9. **B**   The sum of the exterior angles of a convex polygon equals $360°$. Therefore $p + r + t + v + x = 360°$. Similarly, $q + s + u + w + y = 360°$. Therefore $p + q + r + s + t + u + v + w + x + y = 720°$.

10. C $2^2 \times 3^3 \times 5^5 \times 7^7$ is of the form $2^2 \times$ an odd number. It therefore has the form $4(2n + 1) = 8n + 4$ where n is a positive integer and so leaves a remainder of 4 when divided by 8.

11. D   As the 3 numbers have mean 7, their sum equals $3 \times 7 = 21$. For one of the numbers to be as large as possible the other two numbers must be as small as possible. They must also be different and so must be 1 and 2. Hence the largest possible of the three numbers equals $21 - (1 + 2) = 18$.

12. D   If the ant moves alternately from white square to black square and from black to white, then it will end on a white square after 4 moves. So it must find a way to move from white to white or from black to black. However, there is only one pair of adjacent black squares and only one of white. To reach that pair of black squares, the ant must move to one side then climb up to one of the pair. That uses up 3 moves, and the fourth must take it to the other black square of that pair. Thus the two black squares in that pair are possible end points.

If, instead, the ant uses the white pair, it must first move to one side, then climb up to one of the white pair then across to the other square of that pair. That uses 3 moves. The fourth move can then take it to any of the three adjoining black squares. This gives 6 end squares, but these include the two already identified. So there are just 6 possible end squares which are black.

13. A Three vertical lines have been added to the diagram. These divide the original diagram into

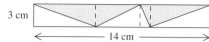

4 rectangles. In each of these, a diagonal divides the rectangle into two triangles of equal area, one shaded and one unshaded. So the total shaded area in the original rectangle is equal to the total unshaded area and is therefore equal to half the area of the original rectangle. So the total shaded area is $\frac{1}{2} \times 3 \times 14 \, \text{cm}^2 = 21 \, \text{cm}^2$.

14. C Suppose the first three terms of the sequence are a, b, c. Then $c = \frac{1}{2}(a + b)$ and so $a + b = 2c$. The mean of the first three terms is then $\frac{1}{3}(a + b + c) = \frac{1}{3}(2c + c) = c$, so the fourth term is c. Similarly, the following terms are all equal to c. Since one of these terms is 26 and $a = 8$ then $b = 2c - a = 52 - 8 = 44$.

15. C The stripes are of equal width, so the width of each stripe is $(72 \div 6) \, \text{cm} = 12 \, \text{cm}$. The diagram shows that the difference between the areas of any two

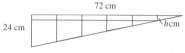

adjacent stripes is equal to the area of a rectangle of width 12 cm and height h cm. By similar triangles, $\frac{h}{12} = \frac{24}{72}$. So $h = \frac{12 \times 24}{72} = 4$. So the required area is $12 \times 4 \, \text{cm}^2 = 48 \, \text{cm}^2$.

16. D All four values cannot be prime. If this were so, both $m \times n$ and $m \div n$ would be prime which can happen only if m is prime and $n = 1$. If m is an odd prime then $m + 1$ is even and at least 4, hence not prime, while if $m = 2$ then $m - 1$ is not prime but $m + 1 = 3$ is. Thus three prime values are the most we can have.

17. D The 12 pentagonal panels have a total of $12 \times 5 = 60$ edges. The 20 hexagonal panels have a total of $20 \times 6 = 120$ edges. So in total the panels have 180 edges. When the panels are sewn together, two edges form each join. So the number of joins is $180 \div 2 = 90$.

18. A Let the weights in kg of the box, 1 plate and 1 cup be b, p and c respectively. Then: $b + 20p + 30c = 4.8$ (i); $b + 40p + 50c = 8.4$ (ii). Subtracting (i) from (ii): $20p + 20c = 3.6$ (iii). So $10p + 10c = 1.8$ (iv). Subtracting (iv) from (i): $b + 10p + 20c = 3$. So the required weight is 3 kg.

19. A The small numbers in the figure show the lengths in cm of each line segment. The larger numbers inside the figure show the areas in cm² of each square or trapezium. (The area of a trapezium is $\frac{1}{2}(a + b)h$ where a and b are the lengths of the parallel sides and h is the perpendicular distance between them.) So the area of the shaded portion in cm² is $11 \times 11 - (1 + 12 + 4 + 15 + 9 + 14 + 16 + 15) = 35$. *(See the extended solutions for a beautifully elegant solution of this problem.)*

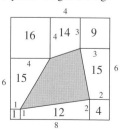

20. C There are 3 different possibilities for the first character. The second character may be any digit from 0 to 9 inclusive, so it has 10 different possibilities. The third character differs from the second digit, so has 9 different possibilities. Once the second and third characters are determined, the fourth character is also determined since it is the units digit of the sum of the second and third characters.
So, the number of different codes is $3 \times 10 \times 9 = 270$.

21. D Let the area of each rectangle be Y. Then the total shaded area is $2(Y - X) + X = 2Y - X$. Therefore $X = \frac{1}{8}(2Y - X)$. So $8X = 2Y - X$, that is $9X = 2Y$. Therefore $\frac{X}{Y} = \frac{2}{9}$.

22. B Let the length of the sides of each small square be x. Then the shaded area is $24x^2$. Let the perimeter of the square be divided into eight line segments, each of length y, and four line segments of length z. Some of these are labelled in the diagram. By Pythagoras' Theorem in triangle ABC: $y^2 + y^2 = (5x)^2$, that is $2y^2 = 25x^2$.

So $y = \frac{5}{\sqrt{2}}x = \frac{5\sqrt{2}}{2}x$. Similarly, in the triangle with hypotenuse CD: $x^2 + x^2 = z^2$, that is $2x^2 = z^2$. So $z = \sqrt{2}x$. Therefore the length of the side of the large square is $2y + z = 5\sqrt{2}x + \sqrt{2}x = 6\sqrt{2}x$. So the area of the large square is $\left(6\sqrt{2}x\right)^2 = 72x^2$. Hence the required fraction is $\frac{24x^2}{72x^2} = \frac{1}{3}$.

23. B The permutations which follow UKIMC in dictionary order are UKMCI, UKMIC, UMCIK, UMCKI, UMICK, UMIKC, UMKCI, UMKIC. There are eight of these, so UKIMC is 112th in the list.

24. E In the diagram V is the point where the perpendicular from P meets TS. Let the side of the square $RSTU$ be x. So the radius of the arc from R to T is x. Therefore SP has length x, PV has length $x - 8$ and VS has length $x - 1$. Applying Pythagoras' Theorem to triangle PVS: $(x - 8)^2 + (x - 1)^2 = x^2$. So $x^2 - 16x + 64 + x^2 - 2x + 1 = x^2$. Therefore $x^2 - 18x + 65 = 0$, so $(x - 5)(x - 13) = 0$. Hence $x = 5$ or $x = 13$, but $x > 8$ so the length of the side of the square $RSTU$ is 13.

25. D Points A, B, C, D, E, F on the perimeter of the triangle are as shown. Let AD have length x so that DB has length $3x$. Let the perpendicular from C to AB have length $4h$. So, by similar triangles, the perpendicular from E to DB has length h. The area of triangle ABC is $\frac{1}{2} \times 4x \times 4h = 8xh$. The area of triangle DBE is $\frac{1}{2} \times 3x \times h = \frac{3}{2}xh$. So the area of triangle DBE is $\frac{3}{16}$ of the area of triangle ABC.

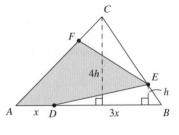

Similarly, by drawing perpendiculars to CB from A and from F, it may be shown that the area of triangle FEC is $\frac{3}{16}$ of the area of triangle ABC.
So the fraction of the area of the triangle that is shaded is $1 - \frac{3}{16} - \frac{3}{16} = \frac{5}{8}$.

The answers

The table below shows the proportion of pupils' choices. The correct answer is shown in bold. [The percentages are rounded to the nearest whole number.]

Qn	A	B	C	D	E	Blank
1	**84**	10	2	1	1	1
2	3	5	5	3	**82**	2
3	3	23	9	9	**52**	4
4	9	6	**40**	20	20	5
5	24	**55**	9	6	2	2
6	6	9	18	**39**	20	8
7	5	34	8	6	**45**	2
8	14	**41**	23	11	5	6
9	13	**38**	18	20	5	6
10	19	13	**26**	18	15	8
11	7	8	8	**47**	24	6
12	44	20	15	**8**	8	4
13	**34**	4	3	3	50	4
14	14	32	**28**	10	5	11
15	7	12	**27**	20	24	10
16	4	6	17	**16**	7	50
17	2	6	15	**25**	10	41
18	**25**	11	9	2	2	50
19	**10**	8	4	5	4	70
20	4	6	**7**	6	8	68
21	2	4	18	**5**	3	67
22	5	**7**	6	4	4	75
23	4	**9**	6	5	5	69
24	3	2	2	6	**5**	82
25	2	2	4	**8**	6	78

IMC 2015: Some comments on the pupils' choice of answers as sent to schools in the letter with the results

The mean score this year of 35 is much lower than last year. The Problems Group would welcome comments from both teachers and pupils on why so many of the questions turned out to be harder than the Problems Group expected. Here we have space only to look at a few of these questions.

It looks as though in Question 3 around three-quarters of the pupils got as far as realising that the fraction they were asked to find is $\frac{1}{6} + \frac{1}{12} + \frac{1}{20}$. This is encouraging, but then nearly a quarter of them simply added the numerators and denominators separately and chose $\frac{3}{38}$ as their answer. You will be able to tell from the table, which compares the answers given by your pupils with the national distribution, how many of your pupils were led in this way to the wrong option B.

Question 4 was intended to be an easy geometry question, so it is disappointing that fewer than half the pupils managed to get it right. We expected pupils to know either that the interior angle of a regular pentagon is 108°, or that its exterior angle is 72°. Were we being too optimistic? The pupils who chose 72° probably knew the exterior angle of a regular pentagon but either failed to think further, or thought that the marked angle was an exterior angle. The choice of 60° by one fifth of the pupils was most likely a guess based on the diagram, and not the result of a reasoned geometrical argument. If many of your pupils did not get this question right, they might benefit by looking at the problems for investigation related to Question 4 to be found in the Extended Solutions on the UKMT website.

In Question 7 it looks as though a third of the pupils did not understand what was meant by *product* and added 4 and 6 rather than multiplying them. If the word product is not used in your classroom to mean the result of multiplying two numbers, and your pupils cannot be expected to know this, please let us know.

It looks as though most pupils, faced in Question 10 with $2^2 \times 3^3 \times 5^5 + 7^7$, an expression of a kind they may not have met before, just guessed. Were we being unreasonable in expecting them to see that this number equals 4 multiplied by an odd number, and so has remainder 4 when divided by 8?

Questions 16 to 25 are harder, and are used to help us select the pupils who are invited to take the Cayley, Hamilton and Maclaurin and Kangaroo papers. Many of these questions are very challenging and most students did not attempt them. Pupils who did well enough on these later, harder questions to qualify for one of the follow-on events should be congratulated on their excellent achievement.

The profile of marks obtained is shown below.

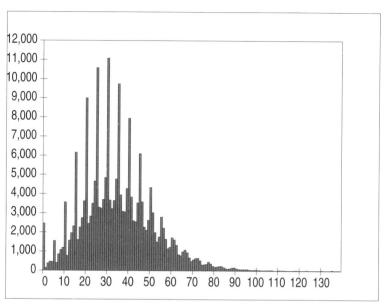

Bar chart showing the actual frequencies in the 2015 IMC

On the basis of the standard proportions used by the UKMT, the cut-off marks were set at

GOLD – 64 or over   SILVER – 49 to 63   BRONZE – 37 to 48

The certificates were virtually identical in design to those used for the JMC.

The cut-off scores for the follow-up competitions were

Year (E&W)	Minimum mark	Event	Minimum mark	Event
11	95	Maclaurin	68	Kangaroo Pink
10	88	Hamilton	68	Kangaroo Pink
9	81	Cayley	61	Kangaroo Grey

The Intermediate Mathematical Olympiad and Kangaroo

(a) *Kangaroo*

The 2015 European Kangaroo (a multiple choice paper with 25 questions) took place on Thursday 19th March. It was also held in many other countries across Europe and beyond with over five million candidates. As in previous years, the UKMT constructed two Kangaroo papers. Invitations were increased in 2015 by 10% to just under 9,000.

EUROPEAN 'KANGAROO' MATHEMATICAL CHALLENGE
'GREY'
Thursday 19th March 2015

Organised by the United Kingdom Mathematics Trust and the Association Kangourou Sans Frontières

This competition is being taken by 6 million students in over 60 countries worldwide.

RULES AND GUIDELINES (to be read before starting):

1. Do not open the paper until the Invigilator tells you to do so.

2. Time allowed: **1 hour**.
   No answers, or personal details, may be entered after the allowed hour is over.

3. The use of rough paper is allowed; **calculators** and measuring instruments are **forbidden**.

4. Candidates in England and Wales must be in School Year 9 or below.
   Candidates in Scotland must be in S2 or below.
   Candidates in Northern Ireland must be in School Year 10 or below.

5. **Use B or HB non-propelling pencil only**. For each question mark *at most one* of the options A, B, C, D, E on the Answer Sheet. Do not mark more than one option.

6. Five marks will be awarded for each correct answer to Questions 1 - 15.
   Six marks will be awarded for each correct answer to Questions 16 - 25.

7. *Do not expect to finish the whole paper in 1 hour.* Concentrate first on Questions 1-15. When you have checked your answers to these, have a go at some of the later questions.

8. The questions on this paper challenge you **to think**, not to guess. Though you will not lose marks for getting answers wrong, you will undoubtedly get more marks, and more satisfaction, by doing a few questions carefully than by guessing lots of answers.

Enquiries about the European Kangaroo should be sent to:
UKMT, School of Mathematics, University of Leeds, Leeds, LS2 9JT.
(Tel. 0113 343 2339)
http://www.ukmt.org.uk

2015 European Grey Kangaroo

1. My umbrella has KANGAROO written on top as shown in the diagram. Which one of the following pictures also shows my umbrella?

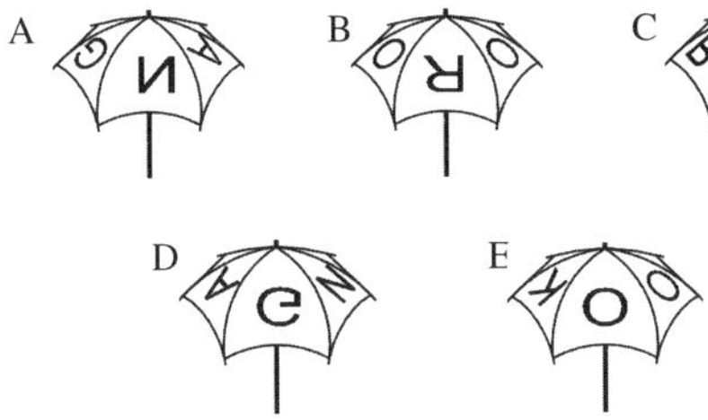

   A   B   C

   D   E

2. Which of the following numbers is closest to 2.015×510.2?

   A 0.1   B 1   C 10   D 100   E 1000

3. Four identical small rectangles are put together to form a large rectangle as shown. The length of a shorter side of each small rectangle is 10 cm. What is the length of a longer side of the large rectangle?

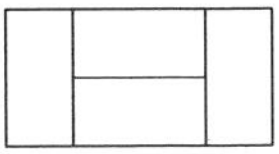

   A 50 cm   B 40 cm   C 30 cm   D 20 cm   E 10 cm

4. Which of the following numbers is not an integer?

   A $\frac{2011}{1}$ B $\frac{2012}{2}$ C $\frac{2013}{3}$ D $\frac{2014}{4}$ E $\frac{2015}{5}$

5. A triangle has sides of lengths 6 cm, 10 cm and 11 cm. An equilateral triangle has the same perimeter. What is the length of the sides of the equilateral triangle?

   A 18 cm   B 11 cm   C 10 cm   D 9 cm   E 6 cm

6. A cyclist rides at 5 metres per second. The wheels of his bicycle have a circumference of 125 cm. How many complete turns does each wheel make in 5 seconds?

   A 4   B 5   C 10   D 20   E 25

7. In a class, no two boys were born on the same day of the week and no two girls were born in the same month. Were another child to join the class, this would no longer be true. How many children are there in the class?

   A 18   B 19   C 20   D 24   E 25

8. In the diagram, the centre of the top square is directly above the common edge of the lower two squares. Each square has sides of length 1 cm. What is the area of the shaded region?

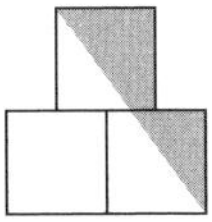

   A $\frac{3}{4}$ cm^2 B $\frac{7}{8}$ cm^2 C 1 cm^2 D $1\frac{1}{4}$ cm^2 E $1\frac{1}{2}$ cm^2

9. Every asterisk in the equation $2 * 0 * 1 * 5 * 2 * 0 * 1 * 5 * 2 * 0 * 1 * 5 = 0$ is to be replaced with either + or − so that the equation is correct. What is the smallest number of asterisks that can be replaced with +?

   A 1   B 2   C 3   D 4   E 5

10. During a rainstorm, 15 litres of water fell per square metre. By how much did the water level in Michael's outdoor pool rise?

   A 150 cm   B 0.15 cm   C 15 cm   D 1.5 cm   E It depends upon the size of the pool

11. A bush has 10 branches. Each branch has either 5 leaves only or 2 leaves and 1 flower. Which of the following could be the total number of leaves the bush has?

   A 45   B 39   C 37   D 31   E None of A to D

12. The mean score of the students who took a mathematics test was 6. Exactly 60% of the students passed the test. The mean score of the students who passed the test was 8. What was the mean score of the students who failed the test?

   A 1   B 2   C 3   D 4   E 5

13. One corner of a square is folded to its centre to form an irregular pentagon as shown in the diagram. The area of the square is 1 unit greater than the area of the pentagon. What is the area of the square?

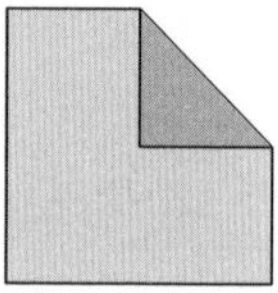

   A 2   B 4   C 8   D 16   E 32

14. Rachel added the lengths of three sides of a rectangle and got 44 cm. Heather added the lengths of three sides of the same rectangle and got 40 cm. What is the perimeter of the rectangle?

   A 42 cm   B 56 cm   C 64 cm   D 84 cm   E 112 cm

15. Luis wants to make a pattern by colouring the sides of the triangles shown in the diagram. He wants each triangle to have one red side, one green side and one blue side. Luis has already coloured some of the sides as shown. What colour can he use for the side marked x?

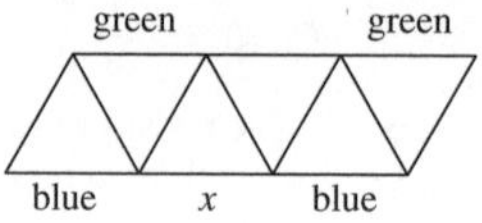

   A only green   B only blue   C only red   D either blue or red
   E The task is impossible

16. Miss Spelling, the English teacher, asked five of her students how many of the five of them had done their homework the day before. Daniel said none, Ellen said only one, Cara said exactly two, Zain said exactly three and Marcus said exactly four. Miss Spelling knew that the students who had not done their homework were not telling the truth but those who had done their homework were telling the truth. How many of these students had done their homework the day before?

   A 0   B 1   C 2   D 3   E 5

17. Ria wants to write a number in each of the seven bounded regions in the diagram. Two regions are neighbours if they share part of their boundary. The number in each region is to be the sum of the numbers in all of its neighbours. Ria has already written in two of the numbers, as shown.

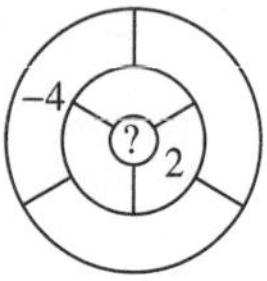

   What number must she write in the central region?

   A 0   B 1   C −2   D −4   E 6

18. Five positive integers (not necessarily all different) are written on five cards. Boris calculates the sum of the numbers on every pair of cards. He obtains only three different totals: 57, 70 and 83. What is the largest integer on any card?

    A 35   B 42   C 48   D 53   E 82

19. A square with area 30 cm² is divided in two by a diagonal and then into triangles as shown.

    The areas of some of these triangles are given in the diagram (which is not drawn to scale). Which part of the diagonal is the longest?

    A *a*   B *b*   C *c*   D *d*   E *e*

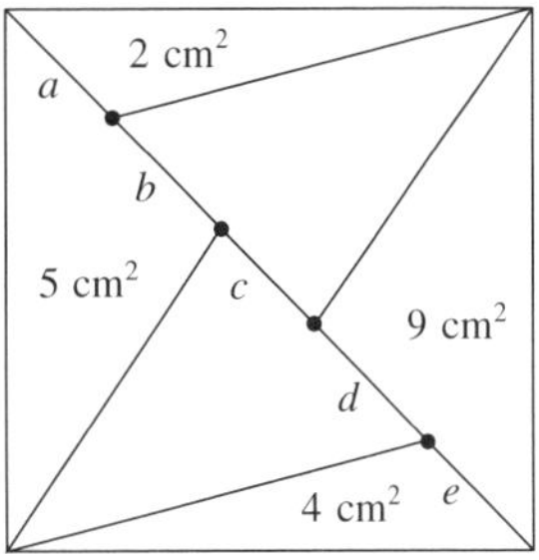

20. In a mob of kangaroos, the two lightest kangaroos together weigh 25% of the total weight of the mob. The three heaviest kangaroos together weigh 60% of the total weight. How many kangaroos are in the mob?

    A 6   B 7   C 8   D 15   E 20

21. Andrew has seven pieces of wire of lengths 1 cm, 2 cm, 3 cm, 4 cm, 5 cm, 6 cm and 7 cm. He bends some of the pieces to form a wire frame in the shape of a cube with edges of length 1 cm without any overlaps. What is the smallest number of these pieces that he can use?

    A 1   B 2   C 3   D 4   E 5

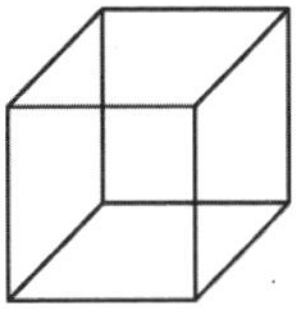

22. In trapezium *PQRS*, the sides *PQ* and *SR* are parallel. Angle *RSP* is 120° and $PS = SR = \frac{1}{3}PQ$. What is the size of angle *PQR*?

    A 15°   B 22.5°   C 25°   D 30°   E 45°

23. Five points lie on a straight line. Alex finds the distances between every pair of points. He obtains, in increasing order, 2, 5, 6, 8, 9, *k*, 15, 17, 20 and 22. What is the value of *k*?

    A 14   B 13   C 12   D 11   E 10

24. Gregor divides 2015 successively by 1, 2, 3, and so on up to and including 1000. He writes down the remainder for each division. What is the largest remainder he writes down?

    A 55   B 215   C 671   D 1007   E some other value

25. Every positive integer is to be coloured according to the following three rules. (i) Each number is to be coloured either red or green. (ii) The sum of any two different red numbers is a red number. (iii) The sum of any two different green numbers is a green number. In how many different ways can this be done?

    A 0   B 2   C 4   D 6   E more than 6

Solutions to the 2015 European Grey Kangaroo

1. **E** In diagrams A, C and D, the letters 'N', 'R' and 'G' respectively have been reversed. In diagram B, the letters are not in the order they appear on the original umbrella. Hence only option E shows part of the original umbrella.

(This is immediately clear if you turn the question paper round so the handles are pointing up rather than down.)

2. **E** Round each number in the product to one significant figure to give $2 \times 500 = 1000$. Hence 1000 is closest to the given product.

3. **B** From the diagram, the length of a small rectangle is twice the width. Hence the length of a small rectangle is 20 cm. Therefore the length of the large rectangle, in cm, is $20 + 2 \times 10 = 40$.

4. **D** The numbers in options A, B and E are clearly integers. In option C, $2 + 0 + 1 + 3 = 6$ so, using the divisibility rule for divisibility by 3, $\dfrac{2013}{3}$ is also an integer. However, while 2000 is divisible by 4, 14 is not so only $\dfrac{2014}{4}$ is not an integer.

5. **D** The perimeter of the equilateral triangle is $(6 + 10 + 11)$ cm $= 27$ cm. Hence the length of the sides of the equilateral triangle is 27 cm $\div 3 = 9$ cm.

6. **D** In five seconds, the cyclist will have travelled 5×5 m $= 25$ m. Hence the wheels will have made $25 \div 1.25 = 20$ complete turns.

7. **B** Suppose another child were to join the class. The question tells us that then one of the two conditions would no longer be true. For this to happen, there must be no day of the week available for a new boy to have been born on and no month of the year available for a new girl to have been born in. Hence there must be 7 boys and 12 girls currently in the class and so there are 19 children in total in the class.

8. **C** The centre of the top square is directly above the common edge of the lower two squares. Hence a rectangle half the size of the square, and so of area $\frac{1}{2}$ cm², can be added to the diagram to form a right-angled triangle as shown. The area of the shaded region and the added rectangle is equal to $\left(\frac{1}{2} \times 2 \times 1\frac{1}{2}\right)$ cm² $= 1\frac{1}{2}$ cm². Hence the area of the shaded region, in cm², is $1\frac{1}{2} - \frac{1}{2} = 1$.

9. **B** The sum of the digits on the left-hand side of the equation is 24. Hence the equation formed by inserting + and − signs must be equivalent to $12 - 12 = 0$. The smallest number of the given digits required to make 12 is three $(2 + 5 + 5)$. Therefore the smallest number of asterisks that can be replaced by + is two and they would be placed in front of two of the three 5s.

10. **D** One litre is equivalent to 1000 cm³. Hence 15 litres falling over an area of one square metre is equivalent to 15 000 cm³ falling over an area of 10 000 cm². Therefore the amount the water in Michael's pool would rise by, in cm, is 15 000 $\div$ 10 000 $= 1.5$.

11. **E** The maximum number of leaves the bush could have is $10 \times 5 = 50$. Each branch that has two leaves and a flower instead of five leaves reduces the number of leaves the bush has by three. Therefore the total number of leaves the bush has is of the form $50 - 3n$ where n is the number of branches with two leaves and a flower. It is straightforward to check that none of options A to D has that form and so the answer is E.

12. C Let the mean score of the 40% of the students who failed the test be x. The information in the question tells us that $0.6 \times 8 + 0.4 \times x = 6$. Hence $0.4x = 1.2$ and so $x = 3$.

13. C The area of the darker triangle in the diagram is $\frac{1}{8}$ of the area of the whole square and this also represents the difference between the area of the square and the area of the pentagon. Hence $\frac{1}{8}$ of the area of the square is equal to 1 unit and so the area of the whole square is 8 units.

14. B Let the length of the rectangle be x cm and let the width be y cm. The information in the question tells us that $2x + y = 44$ and $x + 2y = 40$. Add these two equations to obtain $3x + 3y = 84$. Hence $x + y = 28$ and so the perimeter, which is equal to $2(x + y)$ cm, is 56 cm.

15. A

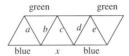

Label the internal sides of the diagram a, b, c, d and e as shown. The side labelled a is in a triangle with a green side and in a triangle with a blue side and so is to be coloured red. This is also the case for the side labelled e. Hence, the side labelled b is in a triangle with a red side and a green side and so is to be coloured blue while the side labelled d is in a triangle with a red side and a blue side and so is to be coloured green. Finally, the side labelled c is in a triangle with a green side and in a triangle with a blue side and so is to be coloured red. Hence the side labelled x is in a triangle with a side that is to be coloured blue (side b) and with a side that is to be coloured red (side c). Therefore the side labelled x is to be coloured green.

16. B All the students have given different answers to the questions so only one, at most, can be telling the truth. Suppose no student is telling the truth; but then Daniel is telling the truth, contradicting this. Hence exactly one student, Ellen, is telling the truth and so only one student had done their homework.

17. E Let the numbers in the four regions that are neighbours to -4 be a, b, c and d as shown in the diagram. The question tells us that $a + b + c + d = -4$. However, we also know that $a + b + c + d + ? = 2$ and hence $? = 6$. (*Note*: The values $a = d = -4$ and $b = c = 2$ give a complete solution to the problem).

18. C Let the five integers be a, b, c, d and e with $a \leqslant b \leqslant c \leqslant d \leqslant e$. The smallest total is 57, which is an odd number so $b \neq a$. Similarly, the largest total is 83, which is also an odd number so $d \neq e$. Hence we now have $a < b \leqslant c \leqslant d < e$ and $a + b = 57$ and $d + e = 83$. Only one possible total remains and so $b = c = d$ with $c + d = 70$. This gives $c = d\, (= b) = 35$ and therefore e, the largest integer, is $83 - 35 = 48$ (whilst $a = 22$).

64

19. D Label the corners of the square A, B, C and D going anticlockwise from the top left corner. Draw in the lines from each marked point on the diagonal to B and to D. All the triangles with a base on the diagonal and a vertex at B or D have the same perpendicular height. Hence their areas are directly proportional to the length of their bases. The two triangles with e as their base both have area 4 cm². Hence the two triangles with base d both have area 5 cm². Similarly the two triangles with base a have area 2 cm², and so the two triangles with base b have area 3 cm². Finally, since the area of the square is 30 cm², the two triangles with base c have area 1 cm². Since the triangles with largest area are those of area 5 cm², the longest base is that part labelled d.

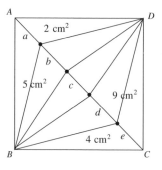

20. A The remaining kangaroos weigh $(100 - 25 - 60)\% = 15\%$ of the total weight. However, this cannot be made up of the weights of more than one kangaroo since the information in the question tells us that the lightest two weigh 25% of the total. Hence there are $2 + 1 + 3 = 6$ kangaroos in the mob.

21. D Since there is no overlap of wires, each vertex of the cube requires at least one end of a piece of wire to form it. A cube has eight vertices and each piece of wire has two ends, so the minimum number of pieces of wire required is $8 \div 2 = 4$.

Such a solution is possible, for example with wires of length 1 cm, 2 cm, 4 cm and 5 cm, the arrangement of which is left for the reader.

22. D

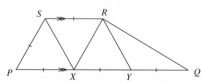

The diagram shows the trapezium described with points X and Y added on PQ so that $PX = XY = YQ = PS = SR$. Since angle $RSP = 120°$, angle $SPQ = 60°$ using co-interior angles and so triangle SPX is equilateral. Similarly, it can easily be shown that triangles SXR and RXY are also equilateral. In triangle RYQ we then have $RY = YQ$ and angle $RYQ = 120°$ using angles on a straight line adding to 180°. Hence triangle RYQ is isosceles and so angle $YQR = \frac{1}{2}(180° - 120°) = 30°$. Therefore angle PQR is 30°.

23. A Let the five points be P, Q, R, S and T with individual distances between them of w, x, y and z as shown.

The maximum distance between any two points is 22 so $PT = 22$. The next largest distance is 20 and, since no distance is 1, this is either PS or QT. Assume $QT = 20$ so that $PQ = w = 2$. The next largest distance is 17 and, since no distance is 3 this cannot be QS or RT and so is PS. Hence $ST = z = 5$ and $QS = 15$. The remaining distances are 6, 8, 9 and k which represent the lengths of PR, QR, RS and RT in some order with $QR + RS = 15$. Since $k > 9$, the only possible pair of distances adding to 15 is 6 and 9 and so $QR = x = 6$ and $RS = y = 9$ (since $QR = 9$ and $RS = 6$ would mean two distances are 11) leaving $PR = 8$ and $RT = k = 14$. Hence the value of k is 14 with the distances between the points taking the values shown below.

(*Note*: The same distances would result but in the reverse order if we assumed $PS = 20$.)

24. C Suppose we obtain the remainder r when we divide 2015 by the positive integer d. Then $r \leqslant d - 1$. Also, with $d \leqslant 1000$ the quotient must be at least 2. This suggests that to get the largest possible remainder we should aim to write 2015 in the form $2d + (d - 1)$. The equation $2015 = 2d + (d - 1)$ has the solution $d = 672$. So we obtain the remainder 671 when we divide 2015 by 672. If we divide 2015 by an integer $d_1 < 672$, the remainder will be at most $d_1 - 1$ and so will be less than 671. If we divide 2015 by an integer d_2, where $672 < d_2 \leqslant 1000$ and obtain remainder r_2, we would have $r_2 = 2015 - 2d_2 < 2015 - 2 \times 672 = 671$. Hence 671 is the largest remainder we can obtain.

25. D There are just six ways to colour the positive integers to meet the two conditions:

    (a) All positive integers are coloured red.

    (b) 1 is coloured red and all the rest are coloured green.

    (c) 1 is coloured red, 2 is coloured green and all the rest are coloured red.

    (d) All positive integers are coloured green.

    (e) 1 is coloured green and all the rest are coloured red.

    (f) 1 is coloured green, 2 is coloured red and all the rest are coloured green.

Note that (d), (e) and (f) can be obtained from (a), (b) and (c) respectively by swapping round the colours red and green.

It is easy to see that these colourings meet the two conditions given. We now need to show that no other colouring does so.

We first show there is no colouring different from a), b) and c) in which the number 1 is coloured red. In such a colouring, since it is different from b), there must be some other number coloured red. Let n be the smallest other number coloured red. Then, as 1 and n are red, so is $n + 1$. Hence $(n + 1) + 1 = n + 2$ is also red, and so on. So all the integers from n onwards are red. Since such a colouring is different from a), $n \neq 2$, and since it is different from c) $n \neq 3$. So $n \geqslant 4$ and therefore 2 and 3 are green. But then $2 + 3 = 5$ is green, so $2 + 5 = 7$ is green and so on. So all positive integers of the form $2k + 3$ are green. In particular $2n + 3$ is green, contradicting the fact that all integers from n onwards are red. Hence no such colouring exists.

A similar argument shows that there is no colouring meeting the given conditions other than (d), (e) and (f) in which 1 is coloured green.

2015 European Pink Kangaroo

1. What is the units digit of the number $2015^2 + 2015^0 + 2015^1 + 2015^5$?

   A 1   B 5   C 6   D 7   E 9

2. The diagram shows a square with sides of length a. The shaded part of the square is bounded by a semicircle and two quarter-circle arcs. What is the shaded area?

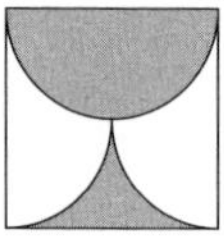

   A $\dfrac{\pi a^2}{8}$ B $\dfrac{a^2}{2}$ C $\dfrac{\pi a^2}{2}$ D $\dfrac{a^2}{4}$ E $\dfrac{\pi a^2}{4}$

3. Mr Hyde can't remember exactly where he has hidden his treasure. He knows it is at least 5 m from his hedge, and at most 5 m from his tree. Which of the following shaded areas could represent the largest region where his treasure could lie?

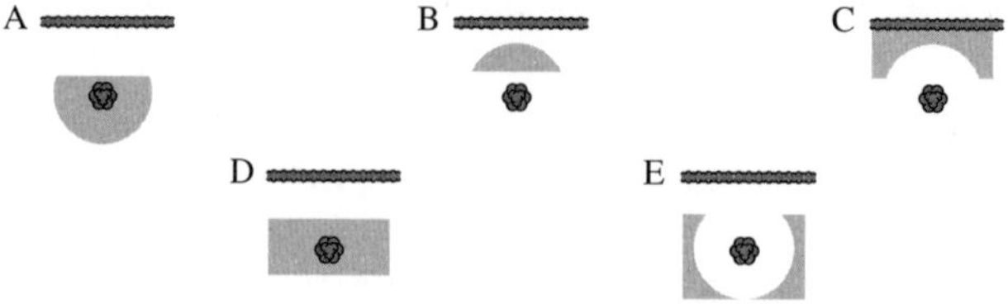

4. Three sisters bought a packet of biscuits for £1.50 and divided them equally among them, each receiving 10 biscuits. However, Anya paid 80 pence, Berini paid 50 pence and Carla paid 20 pence. If the biscuits had been divided in the same ratios as the amounts each sister had paid, how many more biscuits would Anya have received than she did originally?

   A 10   B 9   C 8   D 7   E 6

5. Each of the children in a class of 33 children likes either PE or Computing, and 3 of them like both. The number who like only PE is half as many as like only Computing. How many students like Computing?

   A 15   B 18   C 20   D 22   E 23

6. Which of the following is neither a square nor a cube?

   A 2^9 B 3^{10} C 4^{11} D 5^{12} E 6^{13}

7. Martha draws some pentagons, and counts the number of right-angles in each of her pentagons. No two of her pentagons have the same number of right-angles. Which of the following is the complete list of possible numbers of right-angles that could occur in Martha's pentagons?

   A 1, 2, 3   B 0, 1, 2, 3, 4   C 0, 1, 2, 3   D 0, 1, 2   E 1, 2

8. The picture shows the same die in three different positions. When the die is rolled, what is the probability of rolling a 'YES' ?

   A $\dfrac{1}{3}$ B $\dfrac{1}{2}$ C $\dfrac{5}{9}$ D $\dfrac{2}{3}$ E $\dfrac{5}{6}$

9. In the grid, each small square has side of length 1. What is the minimum distance from 'Start' to 'Finish' travelling only on edges or diagonals of the squares?

A $2\sqrt{2}$ B $\sqrt{10}+\sqrt{2}$ C $2+2\sqrt{2}$ D $4\sqrt{2}$ E 6

10. Three inhabitants of the planet Zog met in a crater and counted each other's ears. Imi said, "I can see exactly 8 ears"; Dimi said, "I can see exactly 7 ears"; Timi said, "I can see exactly 5 ears". None of them could see their own ears. How many ears does Timi have?

A 2   B 4   C 5   D 6   E 7

11. The square $FGHI$ has area 80. Points J, K, L, M are marked on the sides of the square so that $FK = GL = HM = IJ$ and $FK = 3KG$. What is the area of the shaded region?

A 40   B 35   C 30   D 25   E 20

12. The product of the ages of a father and his son is 2015. What is the difference between their ages?

A 29   B 31   C 34   D 36   E None of these

13. A large set of weighing scales has two identical sets of scales placed on it, one on each pan. Four weights W, X, Y, Z are placed on the weighing scales as shown in the left diagram.

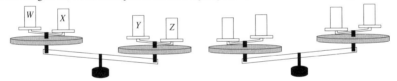

Then two of these weights are swapped, and the pans now appear as shown in the diagram on the right. Which two weights were swapped?

A W and Z B W and Y C W and X D X and Z E X and Y

14. The two roots of the quadratic equation

$$x^2 - 85x + c = 0$$

are both prime numbers. What is the sum of the digits of c?

A 12   B 13   C 14   D 15   E 21

15. How many three-digit numbers are there in which any two adjacent digits differ by 3?

A 12   B 14   C 16   D 18   E 20

16. Which of the following values of n is a counterexample to the statement, 'If n is a prime number, then exactly one of $n - 2$ and $n + 2$ is prime' ?

A 11   B 19   C 21   D 29   E 37

68

17. The figure shows seven regions enclosed by three circles. We call two regions neighbouring if their boundaries have more than one common point. In each region a number is written. The number in any region is equal to the sum of the numbers of its neighbouring regions. Two of the numbers are shown. What number is written in the central region?

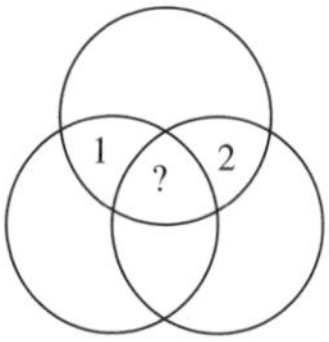

    A –6   B 6   C –3   D 3   E 0

18. Petra has three different dictionaries and two different novels on a shelf. How many ways are there to arrange the books if she wants to keep the dictionaries together and the novels together?

    A 12   B 24   C 30   D 60   E 120

19. How many 2-digit numbers can be written as the sum of exactly six different powers of 2, including 2^0 ?

    A 0   B 1   C 2   D 3   E 4

20. In the triangle FGH, we can draw a line parallel to its base FG, through point X or Y. The areas of the shaded regions are the same. The ratio $HX : XF = 4 : 1$. What is the ratio $HY : YF$?

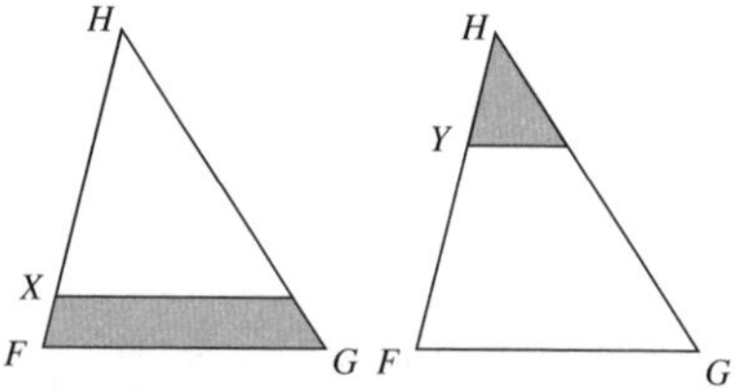

    A 1:1   B 2:1   C 3:1
          D 3:2   E 4:3

21. In a right-angled triangle, the angle bisector of an acute angle divides the opposite side into segments of length 1 and 2. What is the length of the bisector?

    A $\sqrt{2}$ B $\sqrt{3}$ C $\sqrt{4}$ D $\sqrt{5}$ E $\sqrt{6}$

22. We use the notation $\overline{ab}$ for the two-digit number with digits a and b. Let a, b, c be different digits. How many ways can you choose the digits a, b, c such that $\overline{ab} < \overline{bc} < \overline{ca}$?

    A 84   B 96   C 504   D 729   E 1000

23. When one number was removed from the set of positive integers from 1 to n, inclusive, the mean of the remaining numbers was 4.75. What number was eliminated?

    A 5   B 7   C 8   D 9   E impossible to determine

24. Ten different numbers (not necessarily integers) are written down. Any number that is equal to the product of the other nine numbers is then underlined. At most, how many numbers can be underlined?

    A 0   B 1   C 2   D 9   E 10

25. Several different points are marked on a line, and all possible line segments are constructed between pairs of these points. One of these points lies on exactly 80 of these segments (not including any segments of which this point is an endpoint). Another one of these points lies on exactly 90 segments (not including any segments of which it is an endpoint). How many points are marked on the line?

    A 20   B 22   C 80   D 85   E 90

Solutions to the 2015 European Pink Kangaroo

1. **C** The units digits of $2015^2, 2015^0, 2015^1, 2015^5$ are 5, 1, 5, 5, which add to 16. Thus, the units digit of the sum is 6.

2. **B** If the semicircle is cut into two quarter-circles, these can be placed next to the other shaded region to fill up half the square. Hence the shaded area is half of the area of the square, namely $\frac{1}{2}a^2$.

3. **A** Points which are at most 5 m from the tree lie on or inside a circle of radius 5 m with its centre at the tree. However, not all of the inside of the circle will be shaded because the treasure is at least 5 m from the hedge, so we should have an unshaded rectangular strip next to the hedge. This leaves the shaded region in A.

4. **E** Anya pays 80p out of a total of 80p + 50p + 20p = 150p. So if the biscuits had been divided in the same ratio as the payments, she would have received $\frac{80}{150} \times 30 = 16$ biscuits. So she would have received $16 - 10 = 6$ more biscuits.

5. **E** There are three children who like both subjects, leaving 30 children to be shared in the ratio 2:1. Hence 20 like only Computer Science, 10 like only PE, 3 like both. The total number who like Computer Science is $20 + 3 = 23$.

6. **E** Using the index law $a^{mn} = (a^m)^n$, we see $2^9 = (2^3)^3$, a cube; $3^{10} = (3^5)^2$, a square; $4^{11} = (2^2)^{11} = (2^{11})^2$, a square; and $5^{12} = (5^4)^3$, a cube. This leaves 6^{13} which is neither a square nor a cube since 6 is neither a square nor a cube, and 13 is not divisible by 2 nor by 3.

7. **C** The diagrams below show that it is possible to find pentagons with 0, 1, 2, 3 right angles. The angles in a pentagon add to 540°, so with 4 right angles, the fifth angle would be $540° - 4 \times 90° = 180°$, which would make the shape flat at that vertex, thus creating a quadrilateral. Also, a pentagon with 5 right angles is not possible, because they wouldn't add up to 540°.

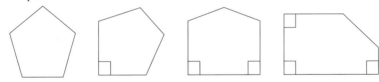

8. **B** We will show that the word "YES" appears exactly three times, giving the probability 3/6 or 1/2. Firstly note that "YES" appears twice on the second die. The third die also shows "YES" and this cannot be the same as either "YES" on the second die: Under the first "YES" is "MAYBE", but on the third die the word "NO" appears below it; to the left of the second "YES" is "MAYBE", but to the left of the "YES" on the third die is "NO". Hence "YES" appears at least three times. However, it appears at most three times because there are two occurrences of "NO" shown in the third die, and one "MAYBE" in the second die. The first die has not been used in the above argument, but is consistent with the faces showing "YES" three times, "NO" twice, and "MAYBE" once.

9. **C** The shortest routes consist of two diagonals (right and down) each of length $\sqrt{2}$, and two sides of length 1, giving a total length $2 + 2\sqrt{2}$.

10. C   The aliens say they can see 8, 7, 5 ears respectively, but each ear has been seen by two aliens so is counted twice. Hence the total number of ears is $\frac{1}{2}(8 + 7 + 5) = 10$ ears. Each alien can see all ten ears, except its own. Timi sees 5 ears, so has $10 - 5 = 5$ ears.

11. D   Let l be the length of KG. Then $FK = 3l$, and the sides of the square $FGHI$ are each $4l$. Since the area of the square is 80, we have $(4l)^2 = 80$, which is $16l^2 = 80$; hence $l^2 = 5$. By Pythagoras' Theorem, $JK^2 = FJ^2 + FK^2 = l^2 + (3l)^2 = 10l^2$. The shaded area is half the area of the square $JKLM$, i.e. half of $JK^2 = \frac{1}{2} \times 10l^2 = 5l^2 = 5 \times 5 = 25$.

12. C   The prime factor decomposition of 2015 is $5 \times 13 \times 31$, so the only possible pairs of ages are 1×2015, 5×403, 13×155, 31×65. The only realistic pair of ages is 65 and 31, with a difference of 34.

13. A   From the first picture, we can see:

From the right scale:	$Z > Y$	(1)
From the left scale:	$X > W$	(2)
From the large scale:	$Y + Z > W + X$	(3)

It follows from (1), (2) and (3) that $Z + Z > Y + Z > W + X > W + W$, so $2Z > 2W$, and hence $Z > W \ldots$ (4).

We can show that most swaps give a contradiction of these inequalities:

Firstly, suppose that weight Z doesn't move. Then there are three possible swaps:

X and Y swap: then in the second picture we must have $Z < X$ and $Y < W$, which add to give $Y + Z < W + X$, contradicting (3).

Y and W swap: then we would have $Z < W$, contradicting (4).

X and W swap: then we would have $X < W$, contradicting (2).

Hence Z must swap, which can happen in three ways:

Z and Y swap: then we would have $Z < Y$, contradicting (1).

Z and X swap: then we would have $Z < W$, contradicting (4).

Z and W swap: This is the only possibility left, and it can work when the weights W, X, Y, Z are 1, 2, 3, 4. The swap would give 4, 2, 3, 1 which matches the picture on the right. However, this does depend on the values of W, X, Y, Z; it would not work for 1, 5, 3, 4.

14. B   Suppose the quadratic has roots p and q. So it factorises as $(x - p)(x - q)$. This expands to give $x^2 - (p + q)x + pq$. Comparing this with $x^2 - 85x + c$ shows that $p + q = 85$ and $pq = c$. It follows that one of p, q is even and one is odd. The only even prime is 2 so the primes p, q are 2 and 83. Therefore $c = 2 \times 83 = 166$, and therefore has digit sum 13.

15. E   The first digit can be anything from 1 to 9. Where possible, we can reduce this by three or increase it by three to get the next digit, giving the following possibilities for the first two digits: 14, 25, 30, 36, 41, 47, 52, 58, 63, 69, 74, 85, 96. Repeating for the third digit, we obtain the possibilities in numerical order: 141, 147, 252, 258, 303, 363, 369, 414, 474, 525, 585, 630, 636, 696, 741, 747, 852, 858, 963, 969, which is 20 options.

16. E Since 21 is not prime, it cannot give a contradiction to the statement because n must be prime.

The primes 11, 19, 29 don't give a contradiction because exactly one of $n - 2, n + 2$ is a prime for each of them: 11 (9 is not prime, 13 is prime), 19 (17 is prime, 21 is not prime), 29 (27 is not prime, 31 is prime).

However for 37, which is prime, this does give a contradiction.

17. E Let x be the number in the central region. Since this is the sum of its three neighbouring regions which include 1 and 2, the region below it must contain $x - 3$. The bottom right region then contains $(x - 3) + 2 = x - 1$. The bottom left region then contains $(x - 3) + 1 = x - 2$. But the number in the bottom central region can now be evaluated in two ways, firstly as $x - 3$, but also as the sum of its neighbours, $x, x - 1, x - 2$. Hence $x - 3 = x + (x - 1) + (x - 2)$, giving $x - 3 = 3x - 3$. So $x = 3x$, giving $2x = 0$ and so $x = 0$.

18. B Petra can arrange the dictionaries in six ways (3 choices for the first dictionary, 2 choices for the second dictionary, 1 choice for the third, giving $3 \times 2 \times 1 = 6$ ways). The novels can be arranged in two ways. Since the novels could be on the left of the dictionaries, or on the right, we have a total of $6 \times 2 \times 2 = 24$ ways to arrange the books.

19. C Since $2^7 = 128$ is larger than 100, the only powers we can choose from are the first seven powers: 2^0 to 2^6, i.e. 1, 2, 4, 8, 16, 32, 64. The sum of all seven is 127. We need to eliminate one of these, and be left with a total under 100, so the only possibilities to remove would be 32 or 64. Hence there are two options:

$1 + 2 + 4 + 8 + 16 + 32 = 63$, and $1 + 2 + 4 + 8 + 16 + 64 = 95$.

20. D In the triangle on the left, the unshaded triangle is similar to triangle FGH, and is obtained from it by a scale factor of $\frac{4}{5}$. Hence its area is $\left(\frac{4}{5}\right)^2 = \frac{16}{25}$ of the area of FGH. The shaded area is therefore $\frac{9}{25}$ of the area of FGH. Hence $HY : HF = 3 : 5$ and so $HY : YF = 3 : 2$.

21. C Label the vertices of the right-angled triangle as A, B, C, with angle $ABC = 90°$. Let M be the point where the angle bisector of CAB meets the side BC. Let N be the point where the perpendicular from M meets the side AC. Let x be the length of the bisector AM.

Then triangles ABM and ANM are congruent (they have all three angles the same, and one corresponding side AM in common). Thus $MN = 1$.

If we reflect triangle MNC in the line NC, then we have an equilateral triangle with all sides equal to length 2. Hence $\angle NCM = 30°$ (half of 60°). Thus $\angle CAB = 180° - 90° - 30° = 60°$, and $\alpha = 30°$. Then triangles MNC and MNA are congruent (all angles the same, and common length MN), so $AM = MC = 2$.

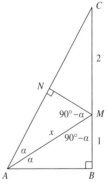

An alternative solution can be obtained using the Angle Bisector Theorem. This gives us $AB : AC = BM : CM = 1 : 2$. Suppose, then, that $AB = x$ and $AC = 2x$. By Pythagoras' Theorem applied to triangle ABC, $(2x)^2 = x^2 + 3^2$. Therefore $x^2 = 3$. Therefore, by Pythagoras' Theorem applied to triangle MBA, $AM^2 = x^2 + 1^2 = 3 + 1 = 4$. Therefore $AM = \sqrt{4} = 2$.

22. A Since the three digits a, b, c are different and $\overline{ab} < \overline{bc} < \overline{ca}$, it is necessary that $a < b < c$. But this condition is also sufficient. Also note that none of a, b, c are zero because they each represent the tens digit of a two-digit number.

There are 9 ways to pick a non-zero digit, 8 ways to pick a second (different) digit, and 7 ways to pick a third digit. These digits can be arranged in $3 \times 2 \times 1 = 6$ ways, but only one of these will be in ascending order (we need $a < b < c$). Hence there are $(9 \times 8 \times 7) \div 6 = 84$ ways to choose the digits a, b, c.

23. B The sum of $1, 2, \ldots n$ is $\frac{1}{2}n(n + 1)$. So the smallest the sum could be if one number were deleted would be $\frac{1}{2}n(n+1) - n = \frac{1}{2}(n^2 + n - 2n) = \frac{1}{2}n(n - 1)$. Thus the mean of any $n - 1$ of these numbers is at least $\frac{1}{2}n$. So for the mean to be 4.75, we must have $n \leqslant 9$.

Similarly, the largest the sum could be with one number deleted would be $\frac{1}{2}n(n + 1) - 1 = \frac{1}{2}(n^2 + n - 2) = \frac{1}{2}(n - 1)(n + 2)$. Thus the mean of any $n - 1$ of the numbers is at most $\frac{1}{2}(n + 2)$. So we must have $\frac{1}{2}(n + 2) \geqslant 4.75$ and so $n \geqslant 8$.

Since $4.75 = 19/4$ and the mean is obtained by dividing by $n - 1$, we require $n - 1$ to be a multiple of 4. So $n = 9$. The sum of $1, \ldots, 9$ is 45 and we need the 8 numbers used to have a mean of 4.75. So their total is 38 and hence 7 is the number to be eliminated.

24. C It is certainly possible to underline two numbers. In the list below both 1 and −1 are equal to the product of the other nine numbers $1, -1, -2, \frac{1}{2}, 3, \frac{1}{3}, 4, \frac{1}{4}, 5, \frac{1}{5}$.

However, it is not possible to underline three numbers. For supposing it was possible to underline the numbers a, b, c. Let N be the product of the other seven numbers. Then we have

$$a = b \times c \times N \ldots (1) \qquad b = a \times c \times N \ldots (2) \qquad c = a \times b \times N \ldots (3)$$

Substituting (1) into (2) gives $b = (b \times c \times N) \times c \times N = b \times c^2 \times N^2$ so $c^2 \times N^2 = 1$. Thus $c \times N = 1$ or $c \times N = -1$. But if $c \times N = 1$, then (1) becomes $a = b$, contradicting that a, b are distinct. Hence $c \times N = -1$, and (1) becomes $a = -b$. Substituting this into (3) gives $c = (-b) \times b \times N$. Multiplying both sides by c gives $c^2 = -b^2 \times c \times N = -b^2 \times (-1) = b^2$. Hence either $c = b$ (contradicting that they are distinct) or $c = -b = a$ (contradicting that a, c are distinct).

Hence there is no way that a, b, c can be distinct numbers and also underlined. (If the numbers are allowed to be the same, it is possible to underline them all by choosing the numbers to be all equal to one).

25. B Suppose the first of the two special points mentioned has a points to its left, and b points to its right. Then the number of line segments it lies on is $a \times b$ (a choices for the left end, b choices for the right end). Also the number of points will be $a + b + 1$. Similarly if the second point has c points to its left, and d to the right, then the number of line segments it lies on is $c \times d$, and the number of points is $c + d + 1$.

Hence we need to find integers a, b, c, d such that $ab = 80$, $cd = 90$, $a + b + 1 = c + d + 1$ (or more simply $a + b = c + d$).

The factor pairs of 80 (and hence the possible values of a, b) are $1 \times 80, 2 \times 40$, $4 \times 20, 5 \times 16, 8 \times 10$.

The factor pairs of 90 (and hence possible values of c, d) are $1 \times 90, 2 \times 45, 3 \times 30$, $5 \times 18, 6 \times 15, 9 \times 10$.

The only pairs for which we have $a + b = c + d$ are 5, 16 and 6, 15. Since both of these add to 21, the number of points must be $a + b + 1 = 21 + 1 = 22$.

(b) *The IMOK Olympiad*

 The United Kingdom Mathematics Trust

Intermediate Mathematical Olympiad and Kangaroo (IMOK)

Olympiad Cayley/Hamilton/Maclaurin Papers

Thursday 19th March 2015

READ THESE INSTRUCTIONS CAREFULLY BEFORE STARTING

1. Time allowed: 2 hours.

2. **The use of calculators, protractors and squared paper is forbidden.**
   Rulers and compasses may be used.

3. Solutions must be written neatly on A4 paper. Sheets must be STAPLED together in the top left corner with the Cover Sheet on top.

4. Start each question on a fresh A4 sheet.
   You may wish to work in rough first, then set out your final solution with clear explanations and proofs. *Do not hand in rough work.*

5. Answers must be FULLY SIMPLIFIED, and EXACT. They may contain symbols such as π, fractions, or square roots, if appropriate, but NOT decimal approximations.

6. Give full written solutions, including mathematical reasons as to why your method is correct.
   Just stating an answer, even a correct one, will earn you very few marks; also, incomplete or poorly presented solutions will not receive full marks.

7. **These problems are meant to be challenging!** The earlier questions tend to be easier; the last two questions are the most demanding.
   Do not hurry, but spend time working carefully on one question before attempting another. Try to finish whole questions even if you cannot do many: you will have done well if you hand in full solutions to two or more questions.

DO NOT OPEN THE PAPER UNTIL INSTRUCTED BY THE INVIGILATOR TO DO SO!

The United Kingdom Mathematics Trust is a Registered Charity.
Enquiries should be sent to: Maths Challenges Office,
School of Mathematics Satellite, University of Leeds, Leeds, LS2 9JT.
(Tel. 0113 343 2339)
http://www.ukmt.org.uk

2015 Olympiad Cayley Paper

> **All candidates must be in** *School Year 9 or below* **(England and Wales),** *S2 or below* **(Scotland), or** *School Year 10 or below* **(Northern Ireland).**

1. A train travelling at constant speed takes five seconds to pass completely through a tunnel which is 85 m long, and eight seconds to pass completely through a second tunnel which is 160 m long.

   What is the speed of the train?

2. The integers a, b, c, d, e, f and g, none of which is negative, satisfy the following five simultaneous equations:

$$a + b + c = 2$$
$$b + c + d = 2$$
$$c + d + e = 2$$
$$d + e + f = 2$$
$$e + f + g = 2.$$

   What is the maximum possible value of $a + b + c + d + e + f + g$?

3. Four straight lines intersect as shown.

   What is the value of $p + q + r + s$?

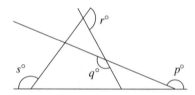

4. Ten balls, each coloured green, red or blue, are placed in a bag. Ten more balls, each coloured green, red or blue, are placed in a second bag.

   In one of the bags there are at least seven blue balls and in the other bag there are at least four red balls. Overall there are half as many green balls as there are blue balls.

   Prove that the total number of red balls in both bags is equal to either the total number of blue balls in both bags or the total number of green balls in both bags.

5. The diagram shows a right-angled triangle and three circles. Each side of the triangle is a diameter of one of the circles. The shaded region R is the region inside the two smaller circles but outside the largest circle.

   Prove that the area of R is equal to the area of the triangle.

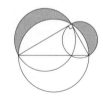

6. I have four identical black beads and four identical white beads.

   Carefully explain how many different bracelets I can make using all the beads.

2015 Olympiad Hamilton Paper

> **All candidates must be in *School Year 10* (England and Wales), *S3* (Scotland), or *School Year 11* (Northern Ireland).**

1. The five-digit integer '$a679b$' is a multiple of 72.

   What are the values of a and b?

2. In Vegetable Village it costs 75 pence to buy 2 potatoes, 3 carrots and 2 leeks at the Outdoor Market, whereas it costs 88 pence to buy 2 potatoes, 3 carrots and 2 leeks at the Indoor Market.

   To buy a potato, a carrot and a leek costs 6 pence more at the Indoor Market than it does at the Outdoor Market.

   What is the difference, in pence, between the cost of buying a carrot at the Outdoor Market and the cost of buying a carrot at the Indoor Market?

3. The diagram shows two circular arcs CP and CQ in a right-angled triangle ABC, where $\angle BCA = 90°$. The centres of the arcs are A and B.

   Prove that $\frac{1}{2}PQ^2 = AP \times BQ$.

   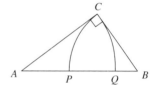

4. The points A, B and C are the centres of three faces of a cuboid that meet at a vertex. The lengths of the sides of the triangle ABC are 4, 5 and 6.

   What is the volume of the cuboid?

5. Some boys and girls are standing in a row, in some order, about to be photographed. All of them are facing the photographer. Each girl counts the number of boys to her left, and each boy counts the number of girls to his right.

   Let the sum of the numbers counted by the girls be G, and the sum of the numbers counted by the boys be B.

   Prove that $G = B$.

6. The diagram shows four identical white triangles symmetrically placed on a grey square. Each triangle is isosceles and right-angled.

   The total area of the visible grey regions is equal to the total area of the white triangles.

   What is the smallest angle in each of the (identical) grey triangles in the diagram?

   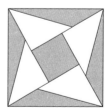

2015 Olympiad Maclaurin Paper

All candidates must be in *School Year 11* (England and Wales), *S4* (Scotland), or *School Year 12* (Northern Ireland).

---

1. Consider the sequence 5, 55, 555, 5555, 55 555, ….

   Are any of the numbers in this sequence divisible by 495; if so, what is the smallest such number?

2. Two real numbers x and y satisfy the equation $x^2 + y^2 + 3xy = 2015$.

   What is the maximum possible value of xy?

3. Two integers are *relatively prime* if their highest common factor is 1.

   I choose six different integers between 90 and 99 inclusive.
   (a) Prove that two of my chosen integers are relatively prime.
   (b) Is it also true that two are *not* relatively prime?

4. The diagram shows two circles with radii a and b and two common tangents AB and PQ. The length of PQ is 14 and the length of AB is 16.

   Prove that $ab = 15$.

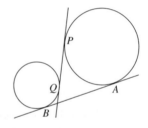

5. Consider equations of the form $ax^2 + bx + c = 0$, where a, b, c are all single-digit prime numbers.

   How many of these equations have at least one solution for x that is an integer?

6. A symmetrical ring of m identical regular n-sided polygons is formed according to the rules:
   (i) each polygon in the ring meets exactly two others;
   (ii) two adjacent polygons have only an edge in common; and
   (iii) the perimeter of the inner region–enclosed by the ring– consists of exactly two edges of each polygon.

   The example in the figure shows a ring with $m = 6$ and $n = 9$.

   For how many different values of n is such a ring possible?

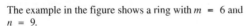

Solutions to the 2015 Olympiad Cayley Paper

C1. A train travelling at constant speed takes five seconds to pass completely through a tunnel which is 85 m long, and eight seconds to pass completely through a second tunnel which is 160 m long. What is the speed of the train?

Solution

Let the train have length l metres.

When the train passes completely through the tunnel, it travels the length of the tunnel plus its own length.

Therefore the first fact tells us that the train travels $85 + l$ metres in five seconds, and the second fact tells us that it travels $160 + l$ metres in eight seconds.

By subtraction, we see that the train travels a distance of $(160 + l) - (85 + l)$ metres in $8 - 5$ seconds, that is, it travels 75 metres in 3 seconds, which corresponds to a speed of 25 m/s.

C2. The integers a, b, c, d, e, f and g, none of which is negative, satisfy the following five simultaneous equations:

$$a + b + c = 2$$
$$b + c + d = 2$$
$$c + d + e = 2$$
$$d + e + f = 2$$
$$e + f + g = 2.$$

What is the maximum possible value of $a + b + c + d + e + f + g$?

Solution

All the equations are satisfied when $a = d = g = 2$ and $b = c = e = f = 0$. In that case we have

$$a + b + c + d + e + f + g = 2 + 0 + 0 + 2 + 0 + 0 + 2$$
$$= 6.$$

Therefore the maximum possible value of $a + b + c + d + e + f + g$ is at least 6. However, from the first and fourth equations we obtain

$$a + b + c + d + e + f + g = (a + b + c) + (d + e + f) + g$$
$$= 2 + 2 + g$$
$$= 4 + g.$$

But the fifth equation is $e + f + g = 2$ and the integers e and f are non-negative, so that g is at most 2. Thus $4 + g$ cannot exceed 6, and thus $a + b + c + d + e + f + g$ cannot exceed 6.

Hence the maximum possible value of $a + b + c + d + e + f + g$ is indeed 6.

Remark

It is possible to use the same method but with different equations in the second part.

C3. Four straight lines intersect as shown.

What is the value of $p + q + r + s$?

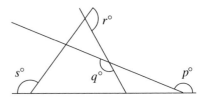

Solution

We present two of the many possible methods of solving this problem.

Method 1

We start by using 'angles on a straight line add up to $180°$' to mark on the diagram three supplementary angles.

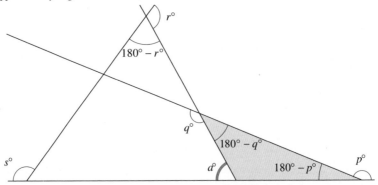

Now we know two of the interior angles in the shaded triangle, and hence can use 'an exterior angle of a triangle is the sum of the two interior opposite angles' to obtain
$a = (180 - p) + (180 - q) = 360 - p - q.$

We now have expressions for two of the interior angles of the triangle shown shaded in the next figure.

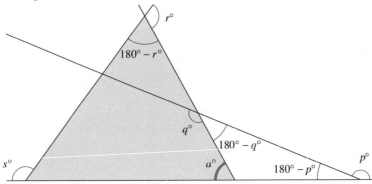

Therefore, once again using 'an exterior angle of a triangle is the sum of the two interior opposite angles', we obtain

$$s = (360 - p - q) + (180 - r),$$

which we may rearrange to give $p + q + r + s = 540$.

Method 2

We start by using 'an exterior angle of a triangle is the sum of the two interior opposite angles' in the two triangles shown shaded in the following figure.

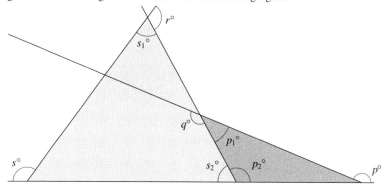

We obtain $s = s_1 + s_2$ and $p = p_1 + p_2$.

Now we notice that we have a pair of angles together at each of three points, so we use 'angles on a straight line add up to $180°$' three times to get

$$p_1 + q = 180$$

$$r + s_1 = 180$$

$$\text{and} \quad p_2 + s_2 = 180.$$

Adding these three results together, we obtain $p_1 + q + r + s_1 + p_2 + s_2 = 540$, from which it follows that $p + q + r + s = 540$.

C4. Ten balls, each coloured green, red or blue, are placed in a bag. Ten more balls, each coloured green, red or blue, are placed in a second bag.

In one of the bags there are at least seven blue balls and in the other bag there are at least four red balls. Overall there are half as many green balls as there are blue balls.

Prove that the total number of red balls in both bags is equal to either the total number of blue balls in both bags or the total number of green balls in both bags.

Solution

Let r, g and b respectively be the numbers of red, green and blue balls that there are in total. From the given information, it follows that $b \geqslant 7$ and $r \geqslant 4$. We also know that $b = 2g$. Since the total number of balls is 20, we have $r + g + b = 20$. Replacing b by $2g$, we see that

$$r = 20 - 3g. \tag{1}$$

We now consider possible values of g in turn.

Case A: $g \leqslant 3$

       In this case $2g \leqslant 6$, so that $b \leqslant 6$ because $b = 2g$. But 6 is less than the specified minimum number of blue balls. So $g \leqslant 3$ is not possible.

Case B: $g = 4$

       In this case $b = 8$ because $b = 2g$, and $r = 8$ from equation (1). Therefore the total number of red balls is equal to the total number of blue balls.

Case C: $g = 5$

       In this case $r = 5$ from equation (1), and $b = 10$. Therefore the total number of red balls is equal to the total number of green balls.

Case D: $g \geqslant 6$

       In this case $20 - 3g \leqslant 2$ from equation (1). Thus $r \leqslant 2$, which is less than the specified minimum number of red balls. So $g \geqslant 6$ is not possible.

Hence in all possible cases the numbers of balls are as we want.

C5. The diagram shows a right-angled triangle and three circles. Each side of the triangle is a diameter of one of the circles. The shaded region R is the region inside the two smaller circles but outside the largest circle.

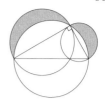

Prove that the area of R is equal to the area of the triangle.

Solution

Let T be the area of the triangle, let S_1 be the area of the semicircle whose diameter is the hypotenuse, and let S_2 and S_3 be the areas of the two semicircles with the shorter sides of the triangle as diameters.

We can view the region in the figure above the hypotenuse in two different ways: either as a semicircle with diameter the hypotenuse plus the shaded region R, or as the triangle plus the two smaller semicircles. This gives us the equation

$$(\text{area of } R) + S_1 = T + S_2 + S_3. \tag{1}$$

Let the hypotenuse of the triangle have length c and let the shorter sides have lengths a and b. Then the semicircles on those sides have radii $\frac{1}{2}c$, $\frac{1}{2}a$ and $\frac{1}{2}b$ respectively. Hence the areas of the semicircles are $S_1 = \frac{1}{8}\pi c^2$, $S_2 = \frac{1}{8}\pi a^2$ and $S_3 = \frac{1}{8}\pi b^2$ respectively. But $a^2 + b^2 = c^2$ by Pythagoras' Theorem, so that we have

$$S_1 = S_2 + S_3. \tag{2}$$

Subtracting equation (2) from equation (1) gives us area of $R = T$, which is what we needed.

C6. I have four identical black beads and four identical white beads.

Carefully explain how many different bracelets I can make using all the beads.

Solution

Firstly, we note the importance of the fact that we are dealing with bracelets: a bracelet can be turned over, as well as turned round. This means that the bracelets shown in the following figures are *not* counted as different.

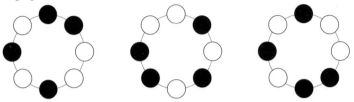

We need to deal with the possibilities in a systematic fashion, to ensure that we count everything.

Our first step is to consider the runs of consecutive black beads. There are five possibilities: $4, 3 + 1, 2 + 2, 2 + 1 + 1$ and $1 + 1 + 1 + 1$. Here the notation $3 + 1$, for example, means that there is a run of three consecutive black beads and a separate run of just one black bead. We consider each case in turn.

Case A: 4

In this case there is only one gap for the white beads, so the four white beads are also consecutive. Just one bracelet is possible, as shown in Figure 1.

Figure 1   Figure 2   Figure 3   Figure 4   Figure 5   Figure 6   Figure 7   Figure 8

Case B: 3 + 1

In this case there are two gaps for the white beads, so there are two possibilities for consecutive runs of white beads: $3 + 1$ or $2 + 2$. Therefore two bracelets are possible, shown in Figure 2 and Figure 3.

Case C: 2 + 2

Once again there are two gaps for the white beads, so there are two possibilities for consecutive runs of white beads: $3 + 1$ or $2 + 2$. Therefore two bracelets are possible, shown in Figure 4 and Figure 5.

Case D: 2 + 1 + 1

In this case there are three gaps for the white beads. So there is only one possibility for consecutive runs of white beads: $2 + 1 + 1$. However, two bracelets are possible, shown in Figure 6 and Figure 7, depending on whether the group of two black beads is adjacent to the group of two white beads, or not.

Case E: 1 + 1 + 1 + 1

Now there are four gaps for the white beads. So there is only one possibility for consecutive runs of white beads: $1 + 1 + 1 + 1$. So only one bracelet is possible, shown in Figure 8.

Therefore exactly eight different bracelets can be made.

Solutions to the 2015 Olympiad Hamilton Paper

H1. The five-digit integer '*a*679*b*' is a multiple of 72.

What are the values of *a* and *b*?

Solution

Since $72 = 8 \times 9$ any number that is a multiple of 72 is both a multiple of 8 and a multiple of 9. Conversely, because 8 and 9 have no common factor greater than 1 and $8 \times 9 = 72$, any number that is both a multiple of 8 and a multiple of 9 is a multiple of 72.

A number is a multiple of 8 if, and only if, its last three digits form a multiple of 8. Hence '79*b*' is a multiple of 8. Since 800 is a multiple of 8, the only number of the required form is 792, and thus $b = 2$.

A number is a multiple of 9 if, and only if, the sum of its digits is a multiple of 9. Hence $a + 6 + 7 + 9 + b$ is a multiple of 9. Because *b* is 2, it follows that $a + 24$ is a multiple of 9, and hence $a = 3$.

Hence $a = 3$ and $b = 2$.

H2. In Vegetable Village it costs 75 pence to buy 2 potatoes, 3 carrots and 2 leeks at the Outdoor Market, whereas it costs 88 pence to buy 2 potatoes, 3 carrots and 2 leeks at the Indoor Market.

To buy a potato, a carrot and a leek costs 6 pence more at the Indoor Market than it does at the Outdoor Market.

What is the difference, in pence, between the cost of buying a carrot at the Outdoor Market and the cost of buying a carrot at the Indoor Market?

Solution

Suppose that a potato costs *p* pence more at the Indoor Market than at the Outdoor Market, that a carrot costs *c* pence more, and that a leek costs *k* pence more. Note that some or all of *p*, *c* and *k* may be negative.

Then from the first sentence we obtain

$$2p + 3c + 2k = 88 - 75$$
$$= 13, \tag{1}$$

and from the second sentence we have

$$p + c + k = 6,$$

so that

$$2p + 2c + 2k = 12. \tag{2}$$

Subtracting equation (2) from equation (1), we get $c = 1$.

Hence the cost of buying a carrot at the Indoor Market is one penny more than buying a carrot at the Outdoor Market.

H3. The diagram shows two circular arcs CP and CQ in a right-angled triangle ABC, where $\angle BCA = 90°$. The centres of the arcs are A and B.

Prove that $\frac{1}{2}PQ^2 = AP \times BQ$.

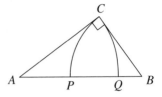

Solution

Let BC, CA and AB have lengths a, b and c respectively. Now triangle ABC is right-angled, so applying Pythagoras' Theorem, we get

$$a^2 + b^2 = c^2.$$

Since A and B are the centres of the arcs, we have $BP = a$ and $AQ = b$, and hence $AP = c - a$ and $BQ = c - b$. Therefore

$$AP \times BQ = (c - a)(c - b)$$
$$= c^2 - ac - bc + ab.$$

Furthermore,

$$PQ = AB - AP - BQ$$
$$= c - (c - a) - (c - b)$$
$$= a + b - c.$$

Hence

$$PQ^2 = (a + b - c)^2$$
$$= a^2 + b^2 + c^2 + 2ab - 2ac - 2bc.$$

So, from $a^2 + b^2 = c^2$, we get

$$PQ^2 = 2c^2 + 2ab - 2ac - 2bc$$
$$= 2 \times AP \times BQ.$$

Thus $\frac{1}{2}PQ^2 = AP \times BQ$, as required.

H4. The points A, B and C are the centres of three faces of a cuboid that meet at a vertex. The lengths of the sides of the triangle ABC are 4, 5 and 6.

What is the volume of the cuboid?

Solution

Let the cuboid have centre O and dimensions $2x \times 2y \times 2z$, as shown. Joining O to each of A, B and C forms three right-angled triangles with sides x, y and 4; y, z and 5; and z, x and 6.

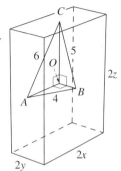

Hence, using Pythagoras' Theorem, we get

$$x^2 + y^2 = 4^2$$

$$y^2 + z^2 = 5^2$$

$$\text{and} \quad z^2 + x^2 = 6^2.$$

Adding all three equations, we obtain

$$2x^2 + 2y^2 + 2z^2 = 16 + 25 + 36$$

$$= 77$$

so that

$$x^2 + y^2 + z^2 = \frac{77}{2}.$$

Subtracting each of the first three equations in turn, we therefore get

$$z^2 = \frac{77}{2} - 16 = \frac{45}{2}$$

$$x^2 = \frac{77}{2} - 25 = \frac{27}{2}$$

$$\text{and} \quad y^2 = \frac{77}{2} - 36 = \frac{5}{2}.$$

Now $2x$ is equal to $\sqrt{4x^2}$ (because x is positive), with similar results for y and z. So the volume of the cuboid $2x \times 2y \times 2z$ is therefore

$$\sqrt{54} \times \sqrt{10} \times \sqrt{90} = \sqrt{54 \times 10 \times 90}$$

$$= \sqrt{6 \times 90 \times 90}$$

$$= 90\sqrt{6}.$$

H5. Some boys and girls are standing in a row, in some order, about to be photographed. All of them are facing the photographer. Each girl counts the number of boys to her left, and each boy counts the number of girls to his right.

Let the sum of the numbers counted by the girls be G, and the sum of the numbers counted by the boys be B.

Prove that $G = B$.

Solution

Consider one of the girls. For each boy to her left, draw an arc connecting the two, as indicated in the following figure (which is shown from the photographer's point of view).

Thus one arc is drawn for each boy that she counts. We repeat this for every girl, and as a result G arcs are drawn altogether.

Now consider adopting the same procedure for the boys. The arcs that would be drawn for the boys are exactly the same as the arcs drawn for the girls. Indeed, in the above figure, the arc corresponds either to a girl with a boy to her left, or to a boy with a girl to his right.

It follows that $G = B$ since both count the total number of arcs.

H6. The diagram shows four identical white triangles symmetrically placed on a grey square. Each triangle is isosceles and right-angled.

The total area of the visible grey regions is equal to the total area of the white triangles.

What is the smallest angle in each of the (identical) grey triangles in the diagram?

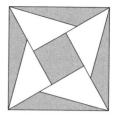

Solution

Choose the unit of measurement so that the two equal sides of each white triangle have length 1. Then the total area of the four white triangles is 2, because they can be assembled to form two squares of side 1.

Therefore the total area of the visible grey regions is also equal to 2, and hence the area of the original grey square is 4. Thus the sides of the grey square have length 2.

Now consider one of the triangles formed by one white triangle and an adjacent grey triangle, such as triangle *ABC* in Figure 1.

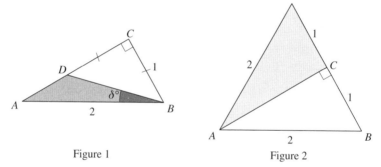

| Figure 1 | Figure 2 |

Figure 1 indicates what we know: the white triangle *BCD* is isosceles and right-angled; the length of *AB* is 2; and the length of *BC* is 1. We are asked to find the value of $\delta°$, which we shall find as the difference between $\angle ABC$ and $\angle DBC$.

Each of the equal angles in an isosceles right-angled triangle is 45° (from 'base angles of an isosceles triangle' and 'the sum of the angles in a triangle is 180°'). Therefore $\angle DBC = 45°$.

In triangle *ABC*, which is right-angled, the length of the hypotenuse is 2, and another side has length 1. It follows that triangle *ABC* is half an equilateral triangle, as shown in Figure 2. But each angle in an equilateral triangle is 60°, so that $\angle ABC = 60°$.

Hence $\delta = 60 - 45$, and therefore the smallest angle in each grey triangle is 15°.

Solutions to the 2015 Olympiad Maclaurin Paper

M1. Consider the sequence 5, 55, 555, 5555, 55 555, ….

Are any of the numbers in this sequence divisible by 495; if so, what is the smallest such number?

Solution

Note that $495 = 5 \times 9 \times 11$, so that a number is divisible by 495 if it is divisible by all of 5, 9 and 11, and not otherwise.

Every number in the sequence has 'units' digit 5, so is divisible by 5.

Each even term of the sequence is divisible by 11, but each odd term is not, since it has a remainder of 5 when divided by 11.

It is therefore enough to find the first even term which is divisible by 9.

Suppose a term has $2k$ digits; then its digit sum is $10k$. But a number is divisible by 9 when its digit sum is divisible by 9, and not otherwise. So the first even term divisible by 9 is the one for which $k = 9$.

Therefore the first term divisible by 495 is the 18th term

$$555\,555\,555\,555\,555\,555,$$

which consists of 18 digits 5.

M2. Two real numbers x and y satisfy the equation $x^2 + y^2 + 3xy = 2015$.

What is the maximum possible value of xy ?

Solution

Subtracting x^2 and y^2 from each side of the given equation, we obtain

$$3xy = 2015 - x^2 - y^2.$$

Now adding $2xy$ to each side, we get

$$5xy = 2015 - x^2 + 2xy - y^2$$
$$= 2015 - (x - y)^2,$$

which has a maximum value of 2015 when $x = y$.

Therefore xy has a maximum value of 403 and this occurs when $x = y = \sqrt{403}$.

M3. Two integers are *relatively prime* if their highest common factor is 1.

I choose six different integers between 90 and 99 inclusive.

   (a) Prove that two of my chosen integers are relatively prime.

   (b) Is it also true that two are *not* relatively prime?

Solution

(a) Consider the five pairs (90, 91), (92, 93), (94, 95), (96, 97), and (98, 99); each is a pair of relatively prime integers. Since we are selecting six numbers, we have to choose two from one pair; thus there is a relatively prime pair.

(b) If two or more of the six numbers chosen are even, then they have a common factor of 2.

If only one is even, then the others are 91, 93, 95, 97 and 99. Therefore two of the chosen numbers (93 and 99) have a common factor of 3.

In either case, two of my six chosen numbers have a common factor greater than 1, so are not relatively prime.

90

M4. The diagram shows two circles with radii a and b and two common tangents AB and PQ. The length of PQ is 14 and the length of AB is 16.

Prove that $ab = 15$.

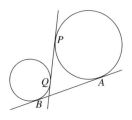

Solution

Let the centres of the circles be C and D.

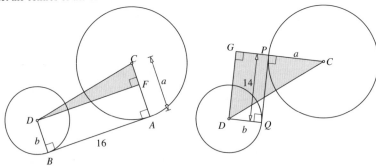

Firstly, consider the tangent AB (see the left-hand figure above). Join C to A and D to B, forming two right angles, as shown, since tangent and radius are perpendicular. Let F be the foot of the perpendicular from D to CA. Then $AFDB$ is a rectangle, so that $AF = b$ and $FD = 16$.

From Pythagoras' Theorem for right-angled triangle CDF, we therefore obtain

$$CD^2 = 16^2 + (a - b)^2. \qquad (1)$$

Next, consider the tangent PQ (see the right-hand figure above). Join C to P and D to Q, once again forming two right angles, as shown, since tangent and radius are perpendicular. Let G be the foot of the perpendicular from D to CP. Then $PGDQ$ is a rectangle, so that $PG = b$ and $GD = 14$.

From Pythagoras' Theorem for right-angled triangle DCG, we therefore obtain

$$CD^2 = 14^2 + (a + b)^2. \qquad (2)$$

It follows from equations (2) and (1) that

$$14^2 + (a + b)^2 = 16^2 + (a - b)^2,$$

and hence

$$(a + b)^2 - (a - b)^2 = 16^2 - 14^2.$$

In order to simplify this equation we could just multiply out the brackets, but it is a little quicker to use the 'difference of two squares' factorisation $x^2 - y^2 = (x - y)(x + y)$ for each side, which also avoids any squaring! On doing this, we get

$$2b \times 2a = 2 \times 30$$

and hence

$$ab = 15.$$

Now in the diagrams we assigned each of a and b to be the radius of a particular circle, whereas the question does not do so. However, interchanging a and b in the result leaves it unchanged, which means that the result is independent of the assignment of letters.

M5. Consider equations of the form $ax^2 + bx + c = 0$, where a, b, c are all single-digit prime numbers.

How many of these equations have at least one solution for x that is an integer?

Solution

Note that a, b and c, being prime, are positive integers greater than 1.

If the equation has an integer solution, then the quadratic expression factorises into two 'linear' brackets $(...)(...)$.

Since c is prime, the 'constant' terms in the brackets can only be c and 1. Similarly, since a is prime, the x terms can only be ax and x. Hence the factorisation is either $(ax + c)(x + 1)$, in which case $b = a + c$, or $(ax + 1)(x + c)$, in which case $b = ac + 1$. Each case leads to an integer solution for x, either $x = -1$ or $x = -c$.

In the first case exactly one of a or c is 2, since b is prime and $b = a + c$. Therefore (a, b, c) is one of $(2, 5, 3)$, $(2, 7, 5)$, $(3, 5, 2)$ or $(5, 7, 2)$.

In the second case one or both of a and c is 2, since b is prime and $b = ac + 1$. Therefore (a, b, c) is one of $(2, 5, 2)$, $(2, 7, 3)$ or $(3, 7, 2)$.

Hence altogether there are seven such quadratic equations.

M6. A symmetrical ring of m identical regular n-sided polygons is formed according to the rules:

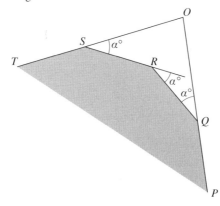

   (i)   each polygon in the ring meets exactly two others;

   (ii)   two adjacent polygons have only an edge in common; and

   (iii)   the perimeter of the inner region–enclosed by the ring– consists of exactly two edges of each polygon.

The example in the figure shows a ring with $m = 6$ and $n = 9$.

For how many different values of n is such a ring possible?

Solution

Let the exterior angle in each polygon be $\alpha°$, so that $\alpha = \dfrac{360}{n}$, and let P, Q, R, S and T be five consecutive vertices of one of the polygons, as shown, where QRS is part of the perimeter of the inner region.

From the symmetry, the point O where PQ and TS meet is the centre of the inner region and hence

$$\angle QOS = \frac{360°}{m}.$$

Now the sum of the angles in quadrilateral $OSRQ$ is $360°$, so we have

$$360 = \frac{360}{m} + \alpha + (180 + \alpha) + \alpha$$

and therefore

$$1 = \frac{2}{m} + \frac{6}{n}.$$

Multiplying each term by mn, we obtain

$$mn = 6m + 2n,$$

so that

$$(m - 2)(n - 6) = 12.$$

Now $m > 2$ for the ring to exist. Hence $(m - 2)(n - 6)$ is a product of two positive integers, so that $n - 6$ is a positive factor of 12. The only possibilities for $n - 6$ are therefore 1, 2, 3, 4, 6 and 12, and thus the only possibilities for n are 7, 8, 9, 10, 12 and 18. The following figures show the corresponding rings.

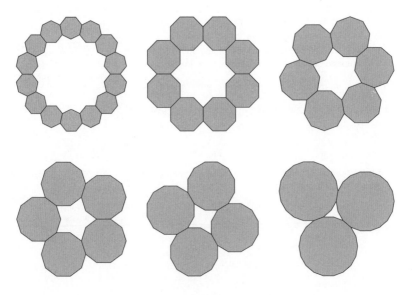

Thus a ring of polygons of the required form is possible for six different values of n.

Comments on the IMOK Olympiad Papers and Scripts

General comments

Both candidates and their teachers will find it helpful to know something of the general principles involved in marking Olympiad-type papers. These preliminary paragraphs therefore serve as an exposition of the 'philosophy' which has guided both the setting and marking of all such papers at all age levels, both nationally and internationally.

What we are looking for, essentially, are solutions to problems. This approach is therefore rather different from what happens in public examinations such as GCSE, AS and A level, where credit is given for the ability to carry out individual techniques regardless of how these techniques fit into a protracted argument. Such marking is cumulative; a candidate may gain 60% of the available marks without necessarily having a clue about how to solve the final problem. Indeed, the questions are generally structured in such a way as to facilitate this approach, divided into many parts and not requiring an overall strategy for tackling a multi-stage argument.

In distinction to this, Olympiad-style problems are marked by looking at each question synoptically and deciding whether the candidate has some sort of overall strategy or not. An answer which is essentially a solution, but might contain either errors of calculation, flaws in logic, omission of cases or technical faults, will be marked on a '10 minus' basis. One question we often ask is: if we were to have the benefit of a two-minute interview with this candidate, could they correct the error or fill the gap? On the other hand, an answer which shows no sign of being a genuine solution is marked on a '0 plus' basis; up to 3 marks might be awarded for particular cases or insights. It is therefore important that candidates taking these papers realise the importance of the rubric about trying to finish whole questions rather than attempting lots of disconnected parts.

Cayley (comments from James Cranch)

IMOK Cayley is the UKMT's toughest individual competition intended specifically for students who are in (or below) Year 9 in England and Wales, Year S2 in Scotland and Year 10 in Northern Ireland. We are now seeing more than six hundred entrants each year: the format is of a single paper with six difficult problems to be attempted over the course of two hours.

In this competition, nothing is intended to be straightforward. Not only is it an achievement to fully solve any question, but the mark schemes are designed so that there are no easy marks: to get even one mark requires a non-trivial idea. We try to ensure that sound thinking will be rewarded better than calculations supported only by guesswork; a correct final answer alone will carry very little weight.

Of course, I was able to personally inspect only a modest fraction of the thousands of individual attempts at problems. But the expert marking team were able to draw my attention to many of the more unusual scripts, and to much interesting mathematics contained within them. As a result of this, here are my comments on this year's paper, question by question.

1. This question contained, as its starting point, a riddle. Several expert mathematicians who attempted the problems in my presence clearly found it far from obvious! The problem, of course, is to explain the claim that the train is travelling at constant speed, given that from a naive perspective the two facts that are given describe different speeds.

   The solution is that, in order to completely pass through a tunnel, a train moves forward not just by the length of the tunnel, but then also by its own length.

   Only once this has been realised can one form equations and solve them to establish the correct answer. Hence a good explanation for the equations is important, since getting this far is the truly unusual part of the problem.

2. This problem, bewildering at first, is (like many such problems) probably best approached by experimentation. Even though such experimentation may end up playing little or no part in the final written solution, it's invaluable for getting to grips with the problem.

   One thing that experimentation can provide is a guess about the correct numerical answer. While it's by no means obvious just from looking at the statement of the problem, it is fairly quick to reach a guess that 6 is the right thing to aim for.

   Another thing that experimentation can help with is the observation that $a = d$, $b = e$, $c = f$ and $d = g$, which is vital for some approaches and helpful for others. This is not particularly difficult from the statement of the problem, but it is experimentation that might suggest it, and also suggest uses.

3. This problem, a classic angle-chase, was known before we started marking to have several different solutions. At the marking weekend. we were delighted to find solutions we had not previously thought of among the many scripts: some involved facts to do with the exterior angles of certain well-chosen polygons, others involved constructing parallel lines to compare various angles in different places.

   Among less good scripts, a common failing was that many students did not document their reasoning intelligibly. Diagrams littered with angles were seen occasionally; unless these diagrams are accompanied by a narrative, it is impossible to tell which angles are determined from which others, and in what order.

4. The key to this problem is probably mostly in organisation: the facts that are given in the problem are hard to relate directly to one another, so one needs to explore their consequences in a tidy and deliberate fashion.

Given the nature of the desired conclusion, it seems likely that the problem will proceed by checking cases in some fashion or other (and that in some cases, the total number of reds will be equal to the total number of blues; in other cases, the total number of reds will be equal to the total number of greens). The challenge is to identify, somehow, a manageable string of cases.

One useful observation, made by the majority of successful candidates, is that considering the bags individually is a needless annoyance. It suffices to work with the totals; and doing so makes the search considerably more straightforward.

5. All known solutions to this problem fall into two distinct parts. One part consists of deploying some technique to manipulate equations for areas, allowing the area R considered in the problem to be compared to something more tractable. The other part involves use of Pythagoras' Theorem to compare semicircles on the two short sides of a right-angled triangle with the one on the hypotenuse.

It was surprising how many students did only the case of a $(3, 4, 5)$-triangle. Not all right-angled triangles are this shape!

6. This question was, as expected, found to be very difficult indeed. As is normal, the main aim is not simply to produce a complete list of all correct bracelets, but to document an analysis that makes it clear that the list that has been produced is correct.

There were a few serious attempts with false claims in them. The usual way of going wrong was to make naive assumptions about symmetry, for example '. . . but we don't care about the difference between a bracelet and its reflection, so we halve the number we counted above'. The problem here is with symmetric bracelets: a bracelet such as ● ○ ○ ● ● ● ○ ○ ● is identical to its reflection, rather than being half of a mirror pair. Counting these is tricky, and the corrections required are even more intricate if one wishes to deal with rotations by dividing out: this is not something that even undergraduates find easy! Nevertheless, some students insisted on attempting it by this method, and a very few were more-or-less correct.

A much better approach is to try to build up the possibilities systematically, based on studying the possible runs of black beads, and how the white beads can fit between them (or vice versa). Having such a system still does not make the problem easy, but it is a good way into it.

Hamilton (comments from Dean Bunnell)

1. This question caused problems for many candidates, with less than half attaining a mark in the '10 minus' category. The main stumbling block was proving that the units digit b equals 2.

    Adopting a 'trial and error' approach could earn pupils full marks, but only if their method was systematic and exhaustive.

    Very few candidates knew that a number is a multiple of 8 if and only if its last three digits form a multiple of 8. Using this method attained the result $b = 2$ very quickly.

    Considering multiples of 4 leads to $b = 2$ or $b = 6$, but many pupils then failed to prove that $b \neq 6$.

    To obtain a mark in the '10 minus' category a candidate had to prove that $b = 2$. Those doing so usually then proceeded correctly to find a.

2. The markers were pleased to see the quality of algebra used by candidates in this admittedly fairly straightforward question.

    Some candidates would have benefitted from using different notation for items of produce at the indoor and outdoor markets. However, as long as it was clear which market the candidate was referring to, a generous approach was taken by the markers.

    A few candidates realised that what really matters is the *difference* in price of each item, and set up the notation accordingly.

3. This popular question was usually approached in a logical manner. The three statements

    $$AC = AQ \text{ (radii)},$$

    $$BC = BP \text{ (radii)},$$

    and $$AB^2 = AC^2 + BC^2 \text{ (Pythagoras' Theorem)},$$

    lead to an equation involving PQ, AP and BQ only. A candidate getting to this stage and no further scored 3 marks.

    To move into the '10 minus' category, candidates had to make a good attempt to form an equation in AP^2, PQ^2, QB^2, $AP \times PQ$, $AP \times QB$ and $PQ \times QB$.

    Candidates quite often neglected to include reasons as to why they had reached a certain conclusion.

4. Some lovely clear diagrams helped pupils' explanations.

    Unfortunately, a few candidates got off to a bad start by taking the length of the sides of the cuboid as 4, 5 and 6, rather than the lengths of the sides of the triangle *ABC*. Whilst they couldn't recover to attain full marks, most managed to reach the '10 minus' category.

The use of Pythagoras' Theorem enabled many candidates to attain at least one dimension of the cuboid correctly. Having made good progress thus far, some failed to form a correct expression for the volume of the cuboid, which may be simplified to $90\sqrt{6}$.

5. There were some very clear explanations of the fact that the sum of the numbers counted by the girls is always equal to the sum of the numbers counted by the boys.

   However, some candidates just considered particular examples, with fixed numbers of girls and boys. In this case the solution was placed in the '0 plus' category.

6. Most candidates used a letter, say x, to represent the equal sides of a white triangle. They then showed that the area of the four white triangles together was equal to $2x^2$, so that the side length of the grey square was equal to $2x$.

   Some solutions used numerical values instead of algebra, which was fine provided the choice of units was made clear at the start.

   To progress from here most candidates considered a right-angled triangle formed by a white triangle and a grey triangle.

   Marks were deducted for a lack of required explanation.

Maclaurin (comments from James Gazet)

There were many very good solutions to these challenging problems and, as always, the markers were impressed by the ingenuity shown by the candidates in producing novel solutions. Many papers were very clearly written up indeed.

It might be useful to summarise some general points before looking at specific questions.

(i) Problems which require extensive searches are seldom if ever set. If a candidate wishes to approach a problem this way, they need to ensure that the search is systematic, correct and complete, else their solution will be penalized heavily.

(ii) It is essential to know whether a problem is about real numbers or about integers only, because on the whole they require different approaches.

(iii) Candidates are advised to always include a diagram in the solution to a geometry problem, otherwise it is difficult for the markers to give credit to solutions which refer to points that are not defined in the question.

(iv) Candidates should know that 1 is not a prime number.

1. This was a popular question. Most candidates approached it by showing that you needed to ensure that the number was divisible by 5, 9 and 11 and using this to limit the number of digits. Divisibility by 5 is immediate and divisibility by 9 required that the number of digits needed to be a multiple of 9: both of these were accepted as statements. Candidates then showed that either the 9-digit number was not divisible by 9 and the 18-digit was, or needed to state that the number of digits needs to be even. Justification either directly or by stating a divisibility rule for 11 completed this proof.

   Some tackled the question by direct calculation, which was awarded full credit provided the details were shown. Some attempted to 'pattern spot' from the remainders when you divide by 495 (they increase by 5, then by 50, then by 5, and so on); failure to justify why this pattern continued was penalized heavily.

2. This was a tricky question. Many candidates wanted either to assume that $x = y$, or to produce an argument to say that $x = y$, and then substitute this directly into the equation. Justifications by symmetry were not rewarded, and some attempts to justify relied on looking at either the $3xy$ term, or the $x^2 + y^2$ terms, without looking at all of them, which is plausible but not correct. The identity $(x + y)^2 + xy = x^2 + y^2 + 3xy$ also appeared frequently, but unfortunately it gives a bound on xy which cannot be achieved.

   A full solution required both the bound and a proof that the bound could be attained. The identity $(x - y)^2 + 5xy = x^2 + y^2 + 3xy$ provides the easiest method. A couple of candidates solved the problem extremely elegantly by applying the AM-GM inequality to the five terms x^2, y^2, xy, xy and xy.

   A few candidates assumed that x and y are integers; sadly, it was difficult to give them any credit.

3. Each part of this question was marked out of 5.

   Successful candidates gave proofs that referred to all possible choices of 6 numbers, rather than a particular choice, or a supposed worst-case scenario. Furthermore, claims that the results were obvious, either from a graph connecting those numbers which were (or were not) relatively coprime, or from tables of factors, needed to ensure they stated what exactly the marker should look for, and that the checking did not require a considerable case study by the marker.

   For part (a), candidates who realized that not many of the numbers shared common factors with more than 4 other numbers frequently came up with succinct proofs. The use of the consecutive integer argument as in the published solutions was relatively rarely seen.

Part (b) was found to be more accessible; several candidates produced the published solution.

4. There were many successful solutions to this approachable question.

Most candidates created two quadrilaterals, one for each circle, with vertices at the points of tangency, the intersection point of the two tangents and the centre of the circle. These two quadrilaterals have the same angles, and it was accepted without penalty that if a candidate made it clear—either from the diagram, or from their write-up—that they appreciated that the two quadrilaterals are kites, they could assert without proof that they are similar. Otherwise, they needed to prove that the quadrilaterals were similar, usually by splitting them up into isosceles or right-angled triangles.

A few candidates approached the proof using Pythagoras' Theorem, as in the published solutions.

5. There was a variety of diverse approaches to this question.

A few candidates tested all 64 possible quadratics; this approach, if done clearly, accurately and with some indication of the method, produced a full mark solution.

Better candidates attacked the problem by limiting the possible values of a, b and c, either by justifying that the expression can be factorised as either $(ax + c)(x + 1)$ or $(ax + 1)(x + c)$, or by limiting the values of b using the discriminant. Both approaches still required some smaller systematic searches to ensure that all the solutions had been found.

Some candidates assumed a, b and c were distinct. This unfortunately reduces rather drastically the number of cases that need to be considered, so a full search based on this was penalized heavily. Those who still attempted some analysis were treated more generously, as were those who thought 1 (and indeed 9) was prime.

6. This question proved to be quite approachable, though time constraints possibly limited the progress of many candidates.

Some first derived the relation $(m - 2)(n - 6) = 12$, by calculating the angle sums in quadrilaterals, relevant m-gons or relevant 2m-gons. From here they finished the problem by looking at the positive factors of 12.

Other candidates showed that, for the configuration to work, $n > 6$, and then proceeded to work out the relevant values of m and n, making it clear that as n is strictly increasing, m is strictly decreasing to its obvious lower bound of 3. Candidates who followed this search approach needed to ensure they made it clear that their search was complete.

A handful of candidates assumed that $\frac{360}{m}$ needed to be an integer, which unfortunately is false.

Marking

The marking was carried out on the weekend of 27th – 29th March in Leeds. There were three marking groups led by James Cranch, Dean Bunnell and James Gazet. The other markers are listed later in this book.

IMOK certificates

All participating students who qualified automatically were awarded a certificate. These came in three varieties: Participation, Merit and Distinction.

THE UKMT INTERMEDIATE MATHEMATICAL OLYMPIAD AND KANGAROO

The IMOK is the follow-on round for the Intermediate Mathematical Challenge and is organised by the UK Mathematics Trust. For each year group, the top scoring 500 or so IMC pupils are invited to participate in the Olympiad, and the next 3000 are invited to participate in the European Kangaroo. Schools may also enter additional pupils to the Olympiad upon payment of a fee; the Kangaroo is by invitation only.

The Olympiad is a two-hour examination which includes six demanding questions requiring full written solutions. The problems are designed to include interesting and attractive mathematics and may involve knowledge or understanding beyond the range of normal school work.

The one-hour multiple choice European Kangaroo requires the use of logic as well as mathematical understanding to solve amusing and thought-provoking questions. The 'Kangourou sans Frontières' is taken by students in over forty countries in Europe and beyond.

The UKMT is a registered educational charity. See our website www.ukmt.org.uk for more information.
Donations would be gratefully received and can be made at
www.donate.ukmt.org.uk if you would like to support our work in this way.

IMOK Olympiad awards

As in recent years, medals were awarded in the Intermediate Mathematical Olympiad. Names of medal winners are listed below. Book prizes were awarded to the top 50 or so in each age group. The Cayley prize was *Seventeen Equations that Changed the World* by Ian Stewart; the Hamilton prize was Fermat's Last Theorem by Simon Singh; and for Maclaurin, *Mathematics: A very short introduction* by Tim Gowers.

The centres of m regular 36-sided polygons, joined edge-to-edge, lie on a circle. How many values of m are possible?

IMOK medal winners

Cayley

Kiran Aberdeen	Queen Elizabeth's School, Barnet
Parth Agarwal	Westminster School
Gianfranco Ameri	Westminster Under School
Pok Aramthanapon	Shrewsbury International School, Thailand
Naomi Bazlov	King Edward VI HS for Girls, Birmingham
Emily Beckford	Lancaster Girls' Grammar School
Oliver Beken	Horndean Technology College, Hampshire
Rose Blyth	Tonbridge Grammar School, Kent
Alexander Buck	St Ninian's High School, Isle of Man
Elena Cates	The Perse School, Cambridge
Justin Chan	Torquay Boys' Grammar School
Dominic Chappell	Redland Green School, Bristol
Richard Chappell	Aylesbury Grammar School
Mark Cheng	Bromsgrove School, Worcestershire
Jinheon Choi	British International School of Shanghai
Soren Choi	Westminster Under School
Gina Choi	North London Collegiate S Jeju, South Korea
Jonathan Coombe	Wilson's School, Surrey
Arianna Cox	Wimbledon High School, Wimbledon
Nathan Creighton	Mossbourne Community Acad, Hackney
Alex Darby	Sutton Grammar School for Boys, Surrey
Stephen Darby	Sutton Grammar School for Boys, Surrey
Timothy De Goede	St James's C of E Secondary S, Bolton
Andy Deng	Wilson's School, Surrey

Kazuki Doi	Yokohama International School
Andrew Dubois	Wellsway School, Bristol
Eamon Dutta Gupta	King Edward VI Grammar S, Chelmsford
Thomas Finn	Bishop Bell C of E School, Eastbourne
Alex Fisher	Reading School
Arul Gupta	Eltham College, London
Trajan Halvorsen	Eton College, Windsor
Freddie Hand	Judd School, Tonbridge, Kent
Jacob Hands	Magdalen College School, Oxford
Josef Hanke	The Cherwell School, Oxford
Faiz Haris Osman	British School of Brussels
Liam Hill	Gosforth Academy, Newcastle-upon-Tyne
Thomas Hillman	St Albans School
Callum Hobbis	Winston Churchill School, Woking
Yang Hsu	St Paul's School, Barnes, London
Dion Huang	Westminster School
Enyu Jin	North London Collegiate S Jeju, South Korea
Stella Johnson	Sch of St Helen and St Katharine, Abingdon
Tushar Kolleri	GEMS Wellington International S, Dubai
Sae Koyama	Rodborough Technology College, Surrey
Soumya Krishna Kumar	Bancroft's School, Essex
Quang Chinh Le	BVIS Ho Chi Minh City
Alex Lee	Taipei European School
Kyung Jae Lee	Garden International School, Malaysia
Qingyang Li	Wychwood School, Oxford
Tony Lin	Wilson's School, Surrey
Mingrui Ma	Ulink College of Suzhou Industrial Park
George Mears	George Abbot School, Guildford
Louis Moen	Tonbridge School, Kent
Isaac Moselle	City of London School
Emre Mutlu	British School of Chicago, Illinois, USA
Hiroharu Nakagawa	Overseas Family School, Singapore
Navonil Neogi	Tiffin School, Kingston-upon-Thames
Hwanjun Noh	North London Collegiate S Jeju, South Korea
Daniel Oosthuizen	The Cherwell School, Oxford
Frank Or	Sha Tin College, Hong Kong
James Panayis	St Albans School

Unnseo Park	International School of Moscow
Tanish Patil	Institut International de Lancy, Switzerland
Frederick Phillips	Aylesbury Grammar School
Elijah Price	Reading School
Chenxin Qiu	Douglas Academy, East Dunbartonshire
David Rae	St Paul's School, Barnes, London
Benedict Randall Shaw	Westminster Under School
Max Rose	Hitchin Boys School, Hertfordshire
Andrea Sendula	Kenilworth School, Warks
Steven Seo	North London Collegiate S Jeju, South Korea
Evans Seow jer ern	Anglo-Chinese School, Singapore
Minhee Shin	North London Collegiate S Jeju, South Korea
Zachary Smith	Beaumont School, St Albans
Alexander Song	Westminster School
Raam Songara	Westminster School
Sam Stansfield	King Edward VI Camp Hill Boys' S, Birmingham
Matthew Strutton	Howard of Effingham School, Surrey
Pasa Suksmith	Harrow School
Adam Thompson	St Paul's School, Barnes, London
Naren Tirumularaju	King Edward's School, Birmingham
Izaak Van Dorgen	Sawston Village College, Cambridgeshire
Tommy Walker Mackay	Stretford Grammar School, Manchester
Anyi Wang	King Edward's School, Birmingham
Sean White	City of London School
Peter Woo	Hampton School, Middlesex
Isaac Wood	Redland Green School, Bristol
Yuqing Wu	Bangkok Patana School
Kelvin Xie	King Edward VI School, Southampton
Bruce Xu	West Island School (ESF), Hong Kong
Christopher Yacoumatos	Eton College, Windsor
Tony Yang	North London Collegiate S Jeju, South Korea
Sang Gon Yi	Garden International School, Malaysia
Eric Yin	Dulwich College Shanghai
Julian Yu	British School Manila, Philippines
Huang Yulong	Beijing New Talent Academy

Hamilton

Julie Ahn	South Island School, Hong Kong
Rahul Arya	King George V School, Hong Kong
Eito Asano	Taunton School (Upper School)
Connie Bambridge-Sutton	Reigate Grammar School, Surrey
Samuel Bealing	Bridgewater High School, Warrington
Shubham Bhargava	UWCSEA East Campus, Singapore
Jonathan Bostock	Eltham College, London
Theo Breeze	Manchester Grammar School
Georgie Bumpus	Meole Brace Science College, Shrewsbury
Gabriel Cairns	Wilson's School, Surrey
Donghyun Chang	South Island School, Hong Kong
Yifei Chen	Graveney School, London
Linxin Chen	Jinan Foreign Language School, China
Alex Chen	Westminster School
Yan Yau Cheng	Discovery College, Hong Kong
Zietto Choi	British International School of Shanghai
George Clements	Norwich School
Matthew Coleclough	Clifton College, Bristol
Louisa Cullen	Pocklington School, nr York
Louise Dai	Cheltenham Ladies' College
Ayman D'Souza	Dulwich College
Andrew Ejemai	Brentwood School
Bertie Ellison-Wright	Bryanston School, Dorset
Reynold Fan	Concord College, Shrewsbury
Mila Feldman	The Stephen Perse Foundation, Cambridge
Xavier Gordon-Brown	Eton College, Windsor
Li Haocheng	Anglo-Chinese School, Singapore
James Hogge	Abingdon School
Katherine Horton	All Hallows Catholic S, Farnham, Surrey
Charlie Hu	City of London School
Clark Huang	Ipswich School
Jake Hyun	Dulwich College Seoul
Isuru Jayasekera	Wilson's School, Surrey
Donghui Jia	King's School Canterbury
Miyamoto Kagetora	Overseas Family School, Singapore
Om Kanchanasakdichai	Winchester College
Minsung Kang	Dulwich College Suzhou, China

Ryan Kang	Westminster School
Basim Khajwal	Heckmondwike Grammar S, West Yorkshire
Chanho Kim	British School of Paris
Eunyoung Kim	Seisen International School, Tokyo
Zion Kim	Hampton School, Middlesex
Minyoung Kim	Claremont Fan Court School, Surrey
Georgina Lang	Moreton Hall School, Oswestry, Shropshire
Timothy Lavy	Magdalen College School, Oxford
Ryan Lee	Magdalen College School, Oxford
Theodore Leebrant	Anglo-Chinese School, Singapore
Ho Ming Leung	Abingdon School
Ziqi Li	Jinan Foreign Language School, China
Ricky Li	Fulford School, York
Yunan Li	Ulink College of Suzhou Industrial Park
Zien Lin	Anglo-Chinese School, Singapore
Aloysius Lip	King Edward's School, Birmingham
Dmitry Lubyako	Eton College, Windsor
Kaiwen Mao	Anglo-Chinese School, Singapore
Laurence Mayther	John Roan School, London
Callum McDougall	Westminster School
Liam McKnight	Magdalen College School, Oxford
Rick Mukhopadhyay	Glasgow Academy
Shuntaro Muto	Deira International School, Dubai
Andy Nam	North London Collegiate S Jeju, South Korea
Karthikeyan Neelamegam	Reading School
Ray Ng	Sha Tin College, Hong Kong
Hong Nhat Nguyen	BVIS Ho Chi Minh City
Dung Nguyen tun	Anglo-Chinese School, Singapore
Shintaro Nishijo	St George's School, Germany
Michael Norman	The Thomas Hardye School, Dorchester
Yuji Okitani	Tapton School, Sheffield
Woohyeok Park	Bromsgrove International School Thailand
Sooyong Park	North London Collegiate S Jeju, South Korea
Jimin Park	Sutton High School, Surrey
Darryl Penson	Focus S, Pulborough Campus, West Sussex
Ryan Po	Island School, Hong Kong
Michae Pung	Anglo-Chinese School, Singapore
Xin Qi	Jinan Foreign Language School, China

Melissa Quail	Longsands Academy, St Neots, Cambs
Lê Quang Trung	BVIS Ho Chi Minh City
Alex Radcliffe	Stewart's Melville College, Edinburgh
William Robson	Eton College, Windsor
Mayuka Saegusa	Henrietta Barnett School, London
Giles Shaw	Bishop Stopford School, Kettering
Andrew Shaw	Verulam School, St Albans
Rohan Shiatis	Weald of Kent Grammar S, Tonbridge
Daniel Starkey	Judd School, Tonbridge, Kent
Morgan Steed	Queen Elizabeth High S, Gainsborough
James Sun	Reading School
Hiroaki Tanioka	Winchester College
Nicholas Tanvis	Anglo-Chinese School, Singapore
Arthur Thomson	Oundle School, Northants
That Tuan Kiet Ton	BVIS Ho Chi Minh City
Daniel Townsend	Colchester Royal Grammar School
Yuriy Tumarkin	Durham Johnston School
Arthur Ushenin	Eton College, Windsor
Laurence Van Someren	Eton College, Windsor
Matthew Wadsworth	Reading School
Ruiyang Wang	Burgess Hill School, West Sussex
Benjamin Wang	Harrow International School, Hong Kong
Yanming Wei	Jinan Foreign Language School, China
Alex Wei	Eton College, Windsor
Naomi Wei	City of London Girls' School
Yannis Wells	Churston Ferrers Grammar School, Devon
Peter Wu	Pocklington School, nr York
Harvey Yau	Ysgol Dyffryn Taf, Carmarthenshire
Harry Yoo	St James School, Grimsby
Hyunji You	Bangkok International Prep and Secondary S
Daniel Yue	King Edward's School, Birmingham
Rodion Zaytsev	King's School Canterbury
Mingqi Zhao	Shrewsbury School
Siana Zhekova	Westminster Academy

Maclaurin

Hugo Aaronson	St Paul's School, Barnes, London
Samuel Ahmed	Kings College School, Wimbledon
John Bamford	The Fernwood School, Nottingham
Thomas Baycroft	Notre Dame High School, Sheffield
Joe Benton	St Paul's School, Barnes, London
Patrick Bevan	The Perse School, Cambridge
Colin Brown	Winchester College
Rosie Cates	The Perse School, Cambridge
Toby Chamberlain	Malmesbury School, Wiltshire
James Chapman	Oundle School, Northants
Jacob Chevalier Drori	Highgate School, London
Tek Kan Chung	Colchester Royal Grammar School
Tudor Ciurca	Chislehurst & Sidcup Grammar S, Kent
Nathaniel Cleland	Gillingham School, Dorset
Timothy Cooper	Myton School, Warwick
Adam Cox	Marling School, Stroud
Jacob Coxon	Magdalen College School, Oxford
Cameron Croucher	Weydon School, Farnham, Surrey
George Cull	Reigate Grammar School, Surrey
William Davies	Reading School
Benjamin Dayan	Westminster School
Monty Evans	St Paul's School, Barnes, London
Wendi Fan	North London Collegiate School
Michael Fernandez	Manchester Grammar School
Alexander Fruh	St Aloysius College, Glasgow
Quentin Gueroult	St Paul's School, Barnes, London
Kaarel Hanni	Southbank International School, London
Edwin Hartanto	Anglo-Chinese School, Singapore
Curtis Ho	Harrow School
Chon Wai Ho	Concord College, Shrewsbury
Alex Horner	City of London School
Xiaoyu Hu	Jinan Foreign Language School, China
Matthew Hutton	Royal Grammar School, Newcastle
Phoebe Jackson	Moreton Hall School, Oswestry, Shropshire
Tiger Ji	Westminster School
Robert Johnson	Aylesbury Grammar School
Toby Jowitt	King Edward's School, Birmingham
Seung Kim	Eton College, Windsor

Jason Kim	St Paul's School, Barnes, London
Kelly Ragyeom Kim	British International School of Shanghai
Minjoo Kim	Frankfurt International School
George Kim	North London Collegiate S Jeju, South Korea
Youngil Ko	British International S Ho Chi Minh City
Rajeev Kumar	Haberdashers' Aske's S for Boys, Herts
William Kusnadi	Anglo-Chinese School, Singapore
Sebastien Laclau	Wellington College, Berkshire
Kyung Chan Lee	Garden International School, Malaysia
Jonathan Lei	South Island School, Hong Kong
Theo Lewy	Haberdashers' Aske's S for Boys, Herts
Alex Li	D'Overbroeck's College, Oxford
Meng Liao	Anglo-Chinese School, Singapore
Anthony Lim	King Edward VI Camp Hill Boys' S, Birmingham
Luozhiyu Lin	Anglo-Chinese School, Singapore
Yuxin Liu	Roedean School, Brighton
Asa MacDermott	Judd School, Tonbridge, Kent
Jacob Mair	Burnham Grammar School, Slough
Diamor Marke	Wallington County Grammar S, Surrey
Rory Mclaurin	Hampstead School, London
Adam Mombru	Eton College, Windsor
Ben Morris	City of London School
Protik Moulik	Westminster School
William Muraszko	The Charter School, London
Michael Ng	Aylesbury Grammar School
Hoang Ngodang	Anglo-Chinese School, Singapore
Nam Nguyen	Anglo-Chinese School, Singapore
Nicholas Palmer	St Paul's School, Barnes, London
Philippe Pangestu	The British International S Jakarta, Indonesia
Matthew Penn	Redland Green School, Bristol
Thomas Pycroft	Whitchurch High School, Cardiff
Otto Pyper	Eton College, Windsor
Mukul Rathi	Nottingham High School
Thomas Read	The Perse School, Cambridge
Matthew Richmond	Cottenham Village College, Cambs
Senan Sekhon	Anglo-Chinese School, Singapore
Jonathan Sewell	Hereford Cathedral School
Joshua Silverbeck	Haberdashers' Aske's S for Boys, Herts
Pratap Singh	The Perse School, Cambridge

Jung Kun Song	Blundell's School, Devon
Timothy Tam	Warwick School
Zoe Tan	St Swithun's School, Winchester
Euan Tebbutt	Twycross House School, Warks
Zarah Tesfai	West Island School (ESF), Hong Kong
Jim Tse	Tiffin School, Kingston-upon-Thames
Yuta Tsuchiya	Queen Elizabeth's School, Barnet
Bhurichaya Tuksinwarajarn	Harrow International School, Bangkok
Sayon Uthayakumar	Bancroft's School, Essex
Alice Vaughan-Williams	Nailsea School, North Somerset
David Veres	King Edward VI School, Southampton
Rohan Virani	Haberdashers' Aske's S for Boys, Herts
Marcus Walford	Kings College School, Wimbledon
Nicholas Wan	Tonbridge School, Kent
Lennie Wells	St Paul's School, Barnes, London
Thomas Wilkinson	Lambeth Academy
Zhen Wu	Anglo-Chinese School, Singapore
Bill Xuan	King Edward's School, Birmingham
Hao Yuan Yang	Mill Hill School, London
Minghua Yin	Reading School
Elliot Young	Wisbech Grammar School
Zhiqiu Yu	Anglo-Chinese School, Singapore
Sechan Yun	The Perse School, Cambridge
Litian Zhang	Jinan Foreign Language School, China
James Zhang	Hutchesons' Grammar School, Glasgow
Yinzi Zhang	Hinchingbrooke S, Huntingdon, Cambs
Yu Xuan Zhou	St Leonard's School, Fife
Liam Zhou	Westminster School

UKMT Summer Schools 2014-2015

Introduction

The first Summer School was held in the Queen's College, Oxford in July 1994. Dr. Tony Gardiner organised and ran the first five summers schools, from 1994 to 1998, when UKMT took over the organisation. From 1997 to 2012 they were held in Queen's College, Birmingham; since then the Trust has organised five annual summer schools, two being held in West Yorkshire, and three in Oxford.

Attendance at the UKMT Summer Schools is by invitation only, and selection of students is based on performance in the UKMT Intermediate Maths Challenge and follow-on rounds.

Summer School for Girls

The Summer School for Girls was held in Oxford from 17th-22nd August 2014. Accommodation was at Balliol College, and teaching was in the Andrew Wiles Building. This year the school was run in partnership with the Clay Mathematics Institute.

Forty students from school years 10 and 11 were invited to attend. There were also 5 senior students who attended to assist the younger girls during the week. These included members of the UK team who went to the European Girls' Mathematical Olympiad.

Oxford Summer Schools

This year two further UKMT Summer Schools were held in Oxford; the first week was 10th − 15th August 2014, and week two 17th − 22nd August 2014. These were both supported by the Department for Education (DfE).

Accommodation for both weeks was at St Anne's College, and teaching was held at the Mathematical Institute (Andrew Wiles Building). Lunches were taken at Somerville College.

Forty students from school years 10 and 11 were invited each week, along with five older students. The senior students had previously attended a summer school and were present to assist and guide the younger pupils as well as to participate in sessions.

The first week was run by Dr Anne Andrews (ex-Royal Latin School, Bucks) and the second week by Philip Coggins (ex-Bedford School).

Summer Schools, West Yorkshire

A further two summer schools were held in July 2015 at Woodhouse Grove School, West Yorkshire. There were 42 students invited to attend each of these weeks and they were assisted each week by six senior students, who once again had been to a previous summer school as a junior and who were on hand to assist the younger pupils.

The first week at Woodhouse Grove was 12th – 17th July 2015 and was run by James Gazet (Eton College). The second week was held between 19th and 24th July and Dr Steven O'Hagan (Hutchesons' Grammar School) was the organiser of this week.

The academic timetables for all UKMT Summer Schools are full and challenging, with masterclasses and group sessions on a host of mathematical topics. However students also enjoy evening activities which include quizzes, talks, a boat or bowling trip and the traditional UKMT summer school concert.

Students attending Summer Schools in 2014-2015

Oxford Summer School (1): Oxford 10th-15th August 2014

Joseph Adams (Knutsford Academy), Jacob Archbold (Benjamin Britten High School), William Bayliff (The Nelson Thomlinson School), Thomas Blackett (Egglescliffe School), Thomas Bowers (All Hallows Catholic High School), Max Campman (Trinity Catholic High School), Edwin Chapman (Wymondham College), Tudor Ciurca (Chislehurst & Sidcup Grammar School), Ben Collard (Kirk Hallam Community Technology College), Ben Collister (Antrim Grammar School), Lily Cooksley (St John the Baptist School), Katie Copeland (Toot Hill School), Adam Cox (Marling School), Robert Davie (Stratton Upper School), Sophie Durrant (Tonbridge Grammar School), Carrie Forsdyke (Wallington High School for Girls), Jamie Forsythe (John Cleveland College), James Goulbourne (Horndean Technology College), Rishabh Gupta (King Edward VI Grammar School), Anthony Hickling (Felsted School), Elizabeth Holdcroft (Willink School), Daisy Hughes (Kingsbridge Community College), Joe Jessener (Willowfield School), Thomas Jones (Macmillan Academy), Phoebe Linane (St Michael's Catholic Grammar School), Tara Madsen (The Tiffin Girls' School), Jacob Mair (Burnham Grammar School), Muhammad Manji (Watford Grammar School for Boys), William Muraszko (The Charter School), Alex Myall (Urmston Grammar School), Tim Parker (Penair School), Alejandro Perez Llabata (St John Fisher High School), Sophia Rogerson (Gryphon School), Gideon Rudolph (Yavneh College), Molly Sayer (Forest School), Dwara Senathirajah (Old Palace

School), Andrew Slattery (Culcheth High School), Francesca Tarrant Cantale (European School), Will Thompson (Elizabeth College), Emma Vinen (Alleyn's School).

Seniors: Jowan Atkinson (Romsey School), Jiali Gao Wolverhampton Girls' High School), Bryony Richards (South Wilts Grammar School), George Robinson (Brooke Weston Academy), Iain Timmin (Wyggeston and Queen Elizabeth I College).

Oxford Summer School (2): 17th – 22nd August 2014

Owen Aljabar (Highdown School), Richard Anderson (George Stephenson High School), George Bateman (Marlborough School), William Briggs (The King's School (Seniors)), Rachel Burton (Olchfa Comprehensive School), Piers Cole (Royal Grammar School, Guildford), Laura Cook (Bishop of Rochester Academy), Dylan Dhokia (Arthur Mellows Village College), Luke Doherty (Cowley International College), Tay Dzonu (Alexandra Park School), Ethan Elstein (University College School), Stewart Feasby (Hazelwick School), Elizabeth Guest (Impington Village College), Jake Hill (Fullbrook School), Edwin Hollands (Queen Elizabeth's Grammar School), Anna Humphreys (Bosworth Academy), Rebecca Hyland (Wheatley Park School), Sarah Jackson (The Crossley Heath School), Robert Zeyu Jin (George Abbot School), Anna Jones (Henrietta Barnett School), Margaret Maxim (The Mountbatten School), Emma Moreby (Central Newcastle High School), Huw Nunn (Bishop of Hereford Bluecoat School), Danielle Parker (Amersham School), Ashley Pearson (Northampton School for Boys), Matthew Penn (Redland Green School), Solene Peroy (Lycee Francais Charles de Gaulle), Emily Richards (Charlton School), James Roper (Upton Court Grammar School), Hannah Samme (Oakgrove School), Daniel Smith (Monmouth School), Jack Terry (Huxlow Science College), Perran Thomas (Portland Place School), Edward Turnbull (Bishop Wordsworth's School), Joanna Ward (Sacred Heart High School), Ethan Webb (The Chase School), Huw Williams (Robert Smyth Academy), Emily Winson-Bushby (Repton School), Ben Woodley (Gregg School), Qingwei Zhang (St Ursula's Convent School).

Seniors: Daniel Clark (Woodhouse Grove School), Sam Gregson (Clitheroe Royal Grammar School), Caroline Harwin (Kendrick School), Robin McCorkell (Dover Grammar School for Boys), Isabel Parsons (Rugby High School).

Summer School for Girls: 17th – 22nd August 2014

Catherine Aldridge (St Philip Howard High School), Rebecca Allen (Stratton Upper School), Holi Ashton (Silverdale School), Anuli Banerjee (Bourne Grammar School), Alice Bennett (Tudor Grange Academy), Verity Bennett (Davison CE High School for Girls), Hannah Bilal (The Crossley Heath School), Hannah Black (The Stephen Perse Foundation), Amy Bradley (Stockport Grammar School), Marsha Burgess (Sheldon School), Rachel Crook (St Bede's Inter-Church School), Hannah Dell (Colyton Grammar School), Manjari Dhar (The Grammar School at Leeds), Wendi Fan (North London Collegiate School), Akane Gonda (Royal Wootton Bassett Academy), Vivien Hasan (Chelmsford County High School), Vera He (Heathfield School), Soraya Hussein (Channing School), Anne Iwashita-Le Roux (Loretto School), Sally Jones (Stanwell School), Rebecca Kennedy (Alfreton Grange Arts College), Elizabeth Knatt (Rodborough Technology College), Weida Liao (Churston Ferrers Grammar School, Katie Lofthouse (Rossett School), Isabel Murray (St Paul's Girls' School), Rachel Onions (Nottingham Girls' High School), Ella Pennington (Withington Girls' School), Imogen Richards (School of St Helen and St Katharine), Megan Richards (Guildford High School), Eilis Rowan (Assumption Grammar School), Mary Scott (Tonbridge Grammar School), Bijal Shah (St Helen's School), Rebecca Shantry (Heathside School), Marina Smith (King Edward's School), Holly Smith (Millais School), Alex Smith (High School of Dundee), Roan Talbut (Perse School), Amy Williams (City of London Girls' School), Molly Williams (Whitchurch High School), Claire Yuanqin Zhang (Leicester High School for Girls).

Seniors: Olivia Aaronson (St. Paul's Girls' School), Penelope Jones (Withington Girls' School), Katya Richards (School of St Helen and St Katharine), Eloise Thuey (Caister Grammar School), Kasia Warburton (Reigate Grammar School).

Woodhouse Grove Summer School (1): 12th – 17th July 2015

Aalia Adam (Henrietta Barnett School), Connie Bambridge-Sutton (Reigate Grammar School), Emily Beatty (King Edward VII School), Patrick Bevan (The Perse School), Jonathan Bostock (Eltham College), Georgie Bumpus (Meole Brace Science College), Yifei Chen (Graveney School), George Clements (Norwich School), Matthew Coleclough (Clifton College), Timothy Cooper (Myton School), Louisa Cullen (Pocklington School), Bertie Ellison-Wright (Bryanston School), Reynold Fan (Concord College), Mila Feldman (The Stephen Perse Foundation), Bethany George (Millfield School), James Hogge (Abingdon School),

Katherine Horton (All Hallows Catholic School), Charlie Hu (City of London School), Phoebe Jackson (Moreton Hall School), Shavindra Isuru Jayasekera (Wilson's School), Donghui Jia (King's School Canterbury), Ryan Joonsuk Kang (Westminster School), Basim Khajwal (Heckmondwike Grammar School), Timothy Lavy (Magdalen College School), Ricky Li (Fulford School), Ritobrata Mukhopadhyay (Glasgow Academy), Karthikeyan Neelamegam (Reading School), Michael Norman (The Thomas Hardye School), Yuji Okitani (Tapton School), Otto Pyper (Eton College), Melissa Quail (Longsands School), Nina Rimsky (City of London Girls' School), Rohan Shiatis (Weald of Kent Grammar School), Rebecca Siddall (Oundle School), Daniel Townsend (Colchester Royal Grammar School), Yuta Tsuchiya (Queen Elizabeth's School), Yuriy Tumarkin (Durham Johnston School), Sayon Uthayakumar (Bancroft's School), Lauren Weaver (St Paul's Girls' School), William Lennie Wells (St Paul's School), Daniel Yue (King Edward's School), James Zhang (Hutchesons' Grammar School).

Seniors: Matthew Chaffe (Littleover Community School), Clarissa Costen (Altrincham Girls' Grammar School), Balaji Krishna (Stanwell School), Georgina Majury (Down High School), Stephen Mitchell (St Paul's School), Sam Watt (Monkton Combe School).

Woodhouse Grove Summer School (2): 19th – 24th July 2015
Joshua Attwell (Chatham & Clarendon Grammar School), Emily Barker (Woodbridge School), Joseph Brason (John F Kennedy School), Isaac Brown (Kingsdale School), Catriona Buck (Queensmead School), James Byrne (Stanborough School), Hazel Cartwright (George Spencer Academy), Mharab Choudhury (Manchester Grammar School), Jonathan Cunningham (Loretto School), Catherine Foster (King's School Ely), Liam Goddard (Saffron Walden County High School), Daniel Hanstock (Chase Terrace Technology College), Thomas Haslam (Dr Challoner's Grammar School), Daniel Hawkins (Brookfield School), Julia Hood (Christ's Hospital), Naomi Hope (Nova Hreod Academy), Sam Jones (Liverpool Blue Coat School), Joris Josiek (Caerleon Comprehensive School), Dominic Littlewood (Dukeries Academy), Joseph Lockhart (Guildford County School), Katya Makukha (St John Lloyd RC Comprehensive School), Shaun Marshall (Shelley College), Tom Martin (Torquay Boys' Grammar School), Nicole Mitchell (Titus Salt School), Aaron Nicoll (Stockport School), Akane Ota (Hall School Wimbledon), Ronan Patel (Swaminarayan School), Dylon Perkins (Bideford College), Kate Pham (King Edward VI Camp Hill Girls' School), Aled Powell (Bitterne Park School), Alex Radcliffe (Stewart's Melville College), Patrick Ramsey

(Chilwell School), Jonti Ruell (Kingsmead School), Aaron Sandhu (Loughborough Grammar School), Gregory Shaw (Hurst Community School), Anya Sims (Headington School), Ryan Skinner (Boston Grammar School), Chris Uren (St Edward's School), Matthew Varley (Wilmington Grammar School for Boys), Robyn Ware (Bridgewater School), Romy Williamson (St Anne's Catholic School).

Seniors: Jowan Atkinson (Romsey School), Meghan Bird (Bourne Grammar School), Harry Ellison-Wright (Bryanston School), Jack Hodkinson (Queen Elizabeth Grammar School), Yifei Painter (Nottingham Girls' High School), Rebecca Poon (George Watson's College).

Our thanks go to everyone who made these Summer Schools such a success, in particular: the Department for Education, the Clay Mathematics Institute, Balliol College, St. Anne's College, Somerville College and the Mathematical Institute, Oxford.

We also thank all our volunteers who worked tirelessly to make these weeks so successful – particularly the leaders of each week – Anne Andrews, Philip Coggins, James Gazet, Lizzie Kimber and Steven O'Hagan. A list of all the volunteers who helped at the summer schools can be found at the back of the Yearbook.

Senior Mathematical Challenge and British Mathematical Olympiads

The Senior Challenge took place on Thursday 6th November 2014, and over 81,000 pupils took part. Once again it was sponsored by the Institute and Faculty of Actuaries. There were 109,660 entries and around 1500 candidates took part in the next stage, British Mathematical Olympiad Round 1, held on Friday 28th November 2014. The Senior Kangaroo was held on the same day, around 3000 candidates were eligible.

UK SENIOR MATHEMATICAL CHALLENGE

Thursday 6 November 2014

Organised by the **United Kingdom Mathematics Trust**

and supported by

Institute
and Faculty
of Actuaries

RULES AND GUIDELINES (to be read before starting)

1. Do not open the question paper until the invigilator tells you to do so.
2. **Use B or HB pencil only**. Mark *at most one* of the options A, B, C, D, E on the Answer Sheet for each question. Do not mark more than one option.
3. Time allowed: **90 minutes**.
   No answers or personal details may be entered on the Answer Sheet after the 90 minutes are over.
4. The use of rough paper is allowed.
   Calculators, measuring instruments and squared paper are forbidden.
5. Candidates must be full-time students at secondary school or FE college, and must be in Year 13 or below (England & Wales); S6 or below (Scotland); Year 14 or below (Northern Ireland).
6. There are twenty-five questions. Each question is followed by five options marked A, B, C, D, E. Only one of these is correct. Enter the letter A-E corresponding to the correct answer in the corresponding box on the Answer Sheet.
7. **Scoring rules**: all candidates start out with 25 marks;
   0 marks are awarded for each question left unanswered;
   4 marks are awarded for each correct answer;
   1 mark is deducted for each incorrect answer.
8. **Guessing**: Remember that there is a penalty for wrong answers. Note also that later questions are deliberately intended to be harder than earlier questions. You are thus advised to concentrate first on solving as many as possible of the first 15-20 questions. Only then should you try later questions.

The United Kingdom Mathematics Trust is a Registered Charity.

http://www.ukmt.org.uk

118

1. What is 98×102?

   A 200 B 9016 C 9996 D 998 E 99 996

2. The diagram shows 6 regions. Each of the regions is to be painted a single colour, so that no two regions sharing an edge have the same colour. What is the smallest number of colours required?

   A 2 B 3 C 4 D 5 E 6

3. December 31st 1997 was a Wednesday. How many Wednesdays were there in 1997?

   A 12 B 51 C 52 D 53 E 365

4. After I had spent $\frac{1}{5}$ of my money and then spent $\frac{1}{4}$ of what was left, I had £15 remaining. How much did I start with?

   A £25 B £75 C £100 D £135 E £300

5. How many integers between 1 and 2014 are multiples of both 20 and 14?

   A 7 B 10 C 14 D 20 E 28

6. In the addition sum shown, each of the letters T, H, I and S represents a non-zero digit.
   What is $T + H + I + S$?

   $$\begin{array}{r} T\ H\ I\ S \\ +\quad I\ S \\ \hline 2\ 0\ 1\ 4 \end{array}$$

   A 34 B 22 C 15 D 9 E 7

7. According to recent research, global sea levels could rise 36.8 cm by the year 2100 as a result of melting ice. Roughly how many millimetres is that per year?

   A 10 B 4 C 1 D 0.4 E 0.1

8. The diagram shows four sets of parallel lines, containing 2, 3, 4 and 5 lines respectively.
   How many points of intersection are there?

   A 54 B 63 C 71 D 95 E 196

9. Which of the following is divisible by 9?

   A $10^{2014} + 5$ B $10^{2014} + 6$ C $10^{2014} + 7$ D $10^{2014} + 8$ E $10^{2014} + 9$

10. A rectangle has area $120\,\text{cm}^2$ and perimeter 46 cm. Which of the following is the length of each of the diagonals?

    A 15 cm B 16 cm C 17 cm D 18 cm E 19 cm

11. A Mersenne prime is a prime of the form $2^p - 1$, where p is also a prime. One of the following is **not** a Mersenne prime. Which one is it?

    A $2^2 - 1$ B $2^3 - 1$ C $2^5 - 1$ D $2^7 - 1$ E $2^{11} - 1$

12. Karen has three times the number of cherries that Lionel has, and twice the number of cherries that Michael has. Michael has seven more cherries than Lionel. How many cherries do Karen, Lionel and Michael have altogether?

    A 12 B 42 C 60 D 77 E 84

13. Each of the five nets P, Q, R, S and T is made from six squares. Both sides of each square have the same colour. Net P is folded to form a cube.

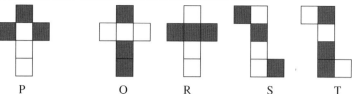

P   Q   R   S   T

How many of the nets Q, R, S and T can be folded to produce a cube that looks the same as that produced by P?

A 0   B 1   C 2   D 3   E 4

14. Given that $\dfrac{3x + y}{x - 3y} = -1$, what is the value of $\dfrac{x + 3y}{3x - y}$?

A −1   B 2   C 4   D 5   E 7

15. The figure shown alongside is made from seven small squares. Some of these squares are to be shaded so that:

    (i) at least two squares are shaded;

    (ii) two squares meeting along an edge or at a corner are not both shaded.

How many ways are there to do this?

A 4   B 8   C 10   D 14   E 18

16. The diagram shows a rectangle measuring 6×12 and a circle.

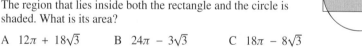

The two shorter sides of the rectangle are tangents to the circle. The circle and rectangle have the same centre.

The region that lies inside both the rectangle and the circle is shaded. What is its area?

A $12\pi + 18\sqrt{3}$ B $24\pi - 3\sqrt{3}$ C $18\pi - 8\sqrt{3}$
         D $18\pi + 12\sqrt{3}$ E $24\pi + 18\sqrt{3}$

17. An oil tanker is 100 km due north of a cruise liner. The tanker sails SE at a speed of 20 kilometres per hour and the liner sails NW at a speed of 10 kilometres per hour. What is the shortest distance between the two boats during the subsequent motion?

A 100km   B 80km   C $50\sqrt{2}$km   D 60km   E $33\frac{1}{3}$km

18. Beatrix decorates the faces of a cube, whose edges have length 2. For each face, she either leaves it blank, or draws a single straight line on it. Every line drawn joins the midpoints of two edges, either opposite or adjacent, as shown.

What is the length of the longest unbroken line that Beatrix can draw on the cube?

A 8   B $4 + 4\sqrt{2}$ C $6 + 3\sqrt{2}$ D $8 + 2\sqrt{2}$ E 12

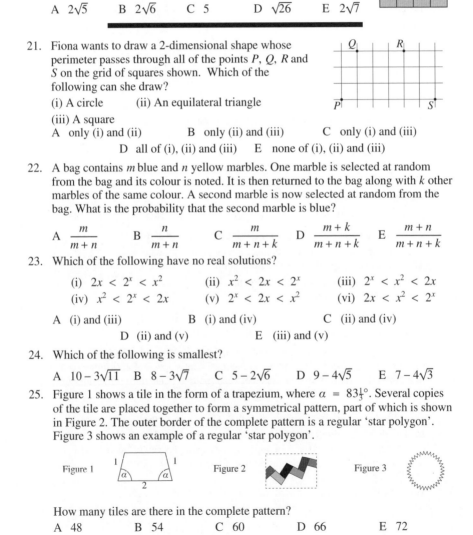

19. The diagram shows a quadrant of radius 2, and two touching semicircles. The larger semicircle has radius 1. What is the radius of the smaller semicircle?

 A $\dfrac{\pi}{6}$ B $\dfrac{\sqrt{3}}{2}$ C $\dfrac{1}{2}$ D $\dfrac{1}{\sqrt{3}}$ E $\dfrac{2}{3}$

20. The diagram shows six squares with sides of length 2 placed edge-to-edge. What is the radius of the smallest circle containing all six squares?

 A $2\sqrt{5}$ B $2\sqrt{6}$ C 5 D $\sqrt{26}$ E $2\sqrt{7}$

21. Fiona wants to draw a 2-dimensional shape whose perimeter passes through all of the points P, Q, R and S on the grid of squares shown. Which of the following can she draw?
 (i) A circle (ii) An equilateral triangle
 (iii) A square
 A only (i) and (ii) B only (ii) and (iii) C only (i) and (iii)
 D all of (i), (ii) and (iii) E none of (i), (ii) and (iii)

22. A bag contains m blue and n yellow marbles. One marble is selected at random from the bag and its colour is noted. It is then returned to the bag along with k other marbles of the same colour. A second marble is now selected at random from the bag. What is the probability that the second marble is blue?

 A $\dfrac{m}{m+n}$ B $\dfrac{n}{m+n}$ C $\dfrac{m}{m+n+k}$ D $\dfrac{m+k}{m+n+k}$ E $\dfrac{m+n}{m+n+k}$

23. Which of the following have no real solutions?
 (i) $2x < 2^x < x^2$ (ii) $x^2 < 2x < 2^x$ (iii) $2^x < x^2 < 2x$
 (iv) $x^2 < 2^x < 2x$ (v) $2^x < 2x < x^2$ (vi) $2x < x^2 < 2^x$
 A (i) and (iii) B (i) and (iv) C (ii) and (iv)
 D (ii) and (v) E (iii) and (v)

24. Which of the following is smallest?
 A $10 - 3\sqrt{11}$ B $8 - 3\sqrt{7}$ C $5 - 2\sqrt{6}$ D $9 - 4\sqrt{5}$ E $7 - 4\sqrt{3}$

25. Figure 1 shows a tile in the form of a trapezium, where $\alpha = 83\frac{1}{3}°$. Several copies of the tile are placed together to form a symmetrical pattern, part of which is shown in Figure 2. The outer border of the complete pattern is a regular 'star polygon'. Figure 3 shows an example of a regular 'star polygon'.

Figure 1 Figure 2 Figure 3

How many tiles are there in the complete pattern?
 A 48 B 54 C 60 D 66 E 72

Further remarks

The solutions are provided.

1.	C
2.	B
3.	D
4.	A
5.	C
6.	B
7.	B
8.	C
9.	D
10.	C
11.	E
12.	D
13.	E
14.	E
15.	C
16.	A
17.	C
18.	D
19.	E
20.	A
21.	B
22.	A
23.	E
24.	A
25.	B

UK SENIOR MATHEMATICAL CHALLENGE

Organised by the **United Kingdom Mathematics Trust**

supported by

Institute
and Faculty
of Actuaries

SOLUTIONS

Keep these solutions secure until after the test on

THURSDAY 6 NOVEMBER 2014

This solutions pamphlet outlines a solution for each problem on this year's paper. We have tried to give the most straightforward approach, but the solutions presented here are not the only possible solutions. Occasionally we have added a 'Note' at the end of a solution.

Please share these solutions with your students.

Much of the potential benefit of grappling with challenging mathematical problems depends on teachers making time for some kind of review, or follow-up, during which students may begin to see what they should have done, and how many problems they could have solved.

We hope that you and they agree that the first 15 problems could, in principle, have been solved by most candidates; if not, please let us know.

The UKMT is a registered charity.

1.   **C** $98 \times 102 = (100 - 2)(100 + 2) = 10000 - 4 = 9996$.

2.   **B**   There are points on the diagram, such as A, where the edges of three regions meet, so three or more different colours are required. A colouring with three colours is possible as shown, so the smallest number of colours required is three.

3.   **D**   The year 1997 was not a leap year so had $365 = 52 \times 7 + 1$ days. Hence, starting from 1st January, 1997 had 52 complete weeks, each starting with the same day as 1st January, followed by 31st December. As 31st December was a Wednesday, so too were all the first days of the 52 complete weeks. So there were 53 Wednesdays in 1997.

4.   **A**   Let the original amount of money be x (in pounds). If I spend $\frac{x}{5}$ then $\frac{4x}{5}$ remains. When I spend $\frac{1}{4}$ of that, $\frac{3}{4}$ of it remains. So $\frac{4x}{5} \times \frac{3}{4}$ is what is left and that is £15. As $\frac{4x}{5} \times \frac{3}{4} = 15$, we have $x = \frac{5}{3} \times 15 = 25$. So the original amount of money is £25.

5.   **C**   The prime factorisations of 20 and 14 are $20 = 2 \times 2 \times 5$ and $14 = 2 \times 7$. The lowest common multiple of 20 and 14 is 140 as $140 = 2 \times 2 \times 5 \times 7$. For a number to be a multiple of 20 and 14 it must be a multiple of 140. As $2014 \div 140 = 14$ remainder 54, there are 14 integers in the required range. Note: The integer 0, which is also a multiple of 20 and of 14 is excluded as we are considering numbers *between* 1 and 2014.

6.   **B**   Working from right to left, the units column shows that $S = 2$ or 7. If $S = 2$, then $I + I = 1$ or 11, neither of which is possible. Hence $S = 7$ and it follows that $I + I = 0$ or 10. However, as the digits are non-zero, $I = 5$. The hundreds column then shows that $H = 9$ and so $T = 1$. This gives $T + H + I + S = 1 + 9 + 5 + 7 = 22$.

7.   **B**   Since $36.8 \div 86$ is approximately $40 \div 100 = 0.4$, the sea level rises by roughly 0.4 cm, which is 4 mm, per year.

8.   **C**   The intersections occur in six groups and the total number of points is $2 \times 3 + 2 \times 4 + 2 \times 5 + 3 \times 4 + 3 \times 5 + 4 \times 5$ which is $6 + 8 + 10 + 12 + 15 + 20 = 71$.

9.   **D**   A number is divisible by 9 if and only if its digit sum is divisible by 9. The number 10^{2014} can be written as a 1 followed by 2014 zeros, so this part has a digit sum of 1. Of all the options given, only adding on 8 to this will make a digit sum of 9, so $10^{2014} + 8$ is the required answer.

10.   **C**   Let the length of the rectangle be x cm and its width be y cm. The area is given as 120 cm^2 so $xy = 120$. The perimeter is 46 cm, so $46 = 2x + 2y$ and therefore $23 = x + y$. Using Pythagoras' Theorem, the length of the diagonal is $\sqrt{x^2 + y^2}$. As $x^2 + y^2 = (x + y)^2 - 2xy$, $\sqrt{x^2 + y^2} = \sqrt{23^2 - 2 \times 120} = \sqrt{529 - 240} = \sqrt{289} = 17$. So the diagonal has length 17 cm.

11.   **E**   First note that the exponent in each of the five options is prime, so we need to see which of the five numbers is not prime. By direct calculation the numbers are 3, 7, 31, 127 and 2047. Only the last number is not prime, as $2047 = 23 \times 89$.

12.   **D**   Let Lionel have x cherries. Michael then has $(x + 7)$ cherries. Karen's number of cherries is described in two ways. She has $3x$ cherries and also $2(x + 7)$ cherries. So $3x = 2x + 14$ and therefore $x = 14$. Lionel has 14 cherries, Michael 21 cherries and Karen 42 cherries giving a total of $14 + 21 + 42 = 77$ cherries.

13.   **E**   Each of P, Q, R, S and T when folded to form a cube consists of a ◰ shape of three black faces and an interlocking ◳ shape of three white faces, so they are all nets of the same cube.

14.   **E**   Rearranging the equation $\dfrac{3x + y}{x - 3y} = -1$ gives $3x + y = -x + 3y$. So $4x = 2y$ and therefore $y = 2x$. Hence $\dfrac{x + 3y}{3x - y} = \dfrac{x + 3 \times 2x}{3x - 2x} = \dfrac{7x}{x} = 7$.

15. C   Label the squares as shown. Possible pairs to be shaded which include A are AD, AE, AF and AG. Pairs excluding A are BD, BG, DE, EG. Triples must include A and there are two possibilities, ADE and AEG. This gives 10 ways of shading the grid,

16. A The diameter of the circle is the same length as the longest sides of the rectangle, so the radius of the circle is 6. The perpendicular distance from the centre of the circle to the longest sides of the rectangle is half of the length of the shortest sides which is 3.

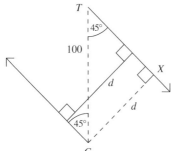

Drawing two diameters AE and DC as shown splits the shaded area into two sectors and two isosceles triangles. As OA is 6 and OB is 3, $\angle AOB = 60°$ and, by Pythagoras' Theorem, $AB = 3\sqrt{3}$. Thus $\angle AOD = 180° - 2 \times 60° = 60°$. So the shaded area is
$$2 \times \frac{60}{360} \times \pi \times 6^2 + 2 \times \frac{1}{2} \times 2 \times 3\sqrt{3} \times 3 = 12\pi + 18\sqrt{3}.$$

17. C The tanker and the cruise liner are travelling in parallel and opposite directions, each making an angle of 45° with the line joining their starting positions. The shortest distance between the ships is d, the perpendicular distance between the parallel lines. This is independent of the speeds of the ships.

Considering triangle TCX gives $\sin 45° = \dfrac{d}{100}$

so $d = \dfrac{1}{\sqrt{2}} \times 100 = 50\sqrt{2}.$

18. D To draw the longest unbroken line Beatrix must be able to draw her design on the net of a cube without taking her pen off the paper. She must minimise the number of lines of length $\sqrt{2}$ and maximise the number of lines of length 2. If no lines of length $\sqrt{2}$ are used, the maximum number of lines of length 2 is four, forming a loop and leaving two faces blank. Thus the longest possible unbroken line would have four lines of length 2 and two lines of length $\sqrt{2}$. A possible configuration to achieve this is shown in the diagram. The length of Beatrix's line is then $8 + 2\sqrt{2}$. Note: This path is not a loop but it is not required to be.

19. E Let the centre of the quadrant be O, the centre of the larger semicircle be A and the centre of the smaller semicircle be B. Let the radius of the smaller semicircle be r. It is given that $OA = 1$. The common tangent to the two semicircles at the point of contact makes an angle of 90° with the radius of each semicircle. Therefore the line AB passes through the point of contact, as $2 \times 90° = 180°$ and angles on a straight line sum to 180°. So the line AB has length $r + 1$. This is the hypotenuse of the right-angled triangle OAB in which $OA = 1$ and $OB = 2 - r$. By Pythagoras' Theorem $(2 - r)^2 + 1^2 = (r + 1)^2$, so $4 - 4r + r^2 + 1 = r^2 + 2r + 1$ and therefore $4 = 6r$ and so $r = \frac{2}{3}.$

20. A It is always possible to draw a circle through three points which are not on a straight line. The smallest circle containing all six squares must pass through (at least) three of the eight vertices of the diagram. Of all such circles, the smallest passes through S, V and Z and has its centre at X. The radius is then $\sqrt{4^2 + 2^2} = \sqrt{16 + 4} = \sqrt{20} = 2\sqrt{5}$.

21. B The diagram shows that it is possible to draw a square whose edges go through P, Q, R and S. By drawing lines through P and S each making an angle of 60° with QR, we can construct an equilateral triangle, as shown, whose edges pass through P, Q, R and S. However there is no circle through these four points. The centre of such a circle would be equidistant from Q and R, and hence would lie on the perpendicular bisector of QR. Similarly it would lie on the perpendicular bisector of PS, but these perpendicular bisectors are parallel lines which don't meet.

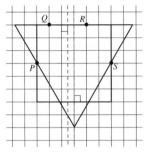

22. A The probability that the second marble is blue equals P(2nd marble is blue given that the 1st marble is blue) + P(2nd marble is blue given that the 1st marble is yellow), which is
$$\frac{m}{m+n} \times \frac{m+k}{m+n+k} + \frac{n}{m+n} \times \frac{m}{m+n+k} = \frac{m^2 + mk + mn}{(m+n)(m+n+k)} = \frac{m(m+k+n)}{(m+n)(m+n+k)} = \frac{m}{m+n}.$$
Note: this expression is independent of k.

23. E If the graphs of $y = 2x$, $y = 2^x$ and $y = x^2$ are sketched on the same axes it can be seen that case (i) holds for $2 < x < 4$, case (ii) holds for $0 < x < 1$, case (iv) holds for $1 < x < 2$ and case (vi) holds for $x > 4$.
There are no real solutions for case (iii). Consider $x^2 < 2x$, which is true for $0 < x < 2$. However for $0 < x < 2$ it can be seen that $2^x > x^2$ rather than $2^x < x^2$ as stated.
There are no real solutions for case (v). Consider $2x < x^2$, which is true for $x < 0$ or $x > 2$. However, when $x < 0$ we have $2^x > 2x$ as 2^x is positive and $2x$ is negative, rather than $2^x < 2x$ as stated. Also, when $x = 2$ we have $2^x = 2x$, but for $x > 2$, $2^x > 2x$ rather than $2^x < 2x$ as stated.

24. A Each of the five expressions can be written in the form $\sqrt{x} - \sqrt{x - 1}$, where x is in turn 100, 64, 25, 81 and 49. As $(\sqrt{x} - \sqrt{x-1})(\sqrt{x} + \sqrt{x-1}) = x - (x - 1) = 1$, we can write $(\sqrt{x} - \sqrt{x-1}) = \dfrac{1}{(\sqrt{x} + \sqrt{x-1})}$. Since $(\sqrt{x} + \sqrt{x-1})$ increases with x, then $(\sqrt{x} - \sqrt{x-1})$ must decrease with x. Therefore, of the given expressions, the one corresponding to the largest value of x is the smallest. This is $\sqrt{100} - \sqrt{99}$ which is $10 - 3\sqrt{11}$.

25. B Let the supplementary angle to α be β. Let tile 1 on the outside of the star polygon be horizontal. Counting anti-clockwise around the star polygon, tile 3 has an angle of elevation from the horizontal of $\beta - \alpha = 96\frac{2}{3}° - 83\frac{1}{3}° = 13\frac{1}{3}°$. As $360° \div 13\frac{1}{3}° = 27$, we need 27 pairs of tiles to complete one revolution. So there are 54 tiles in the complete pattern.

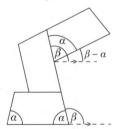

For reasons of space, these solutions are necessarily brief. There are more in-depth, extended solutions available on the UKMT website, which include some exercises for further investigation:
http://www.ukmt.org.uk/

The answers

The table below shows the proportion of pupils' choices. The correct answer is shown in bold. [The percentages are rounded to the nearest whole number.]

Qn	A	B	C	D	E	Blank
1	0	1	**96**	0	1	1
2	1	**57**	34	4	2	1
3	0	10	36	**46**	0	8
4	**69**	4	5	1	19	2
5	12	7	**51**	7	7	17
6	3	**67**	2	13	2	12
7	1	**69**	1	20	2	6
8	2	7	**87**	1	1	2
9	3	2	3	**68**	10	14
10	11	6	**56**	4	3	21
11	8	2	5	19	**49**	16
12	1	11	2	**73**	3	10
13	4	8	10	12	**59**	7
14	19	7	3	2	**44**	25
15	5	31	**26**	12	11	15
16	**13**	8	7	5	3	65
17	11	5	**15**	5	5	58
18	8	12	10	**20**	4	46
19	2	6	5	4	**19**	63
20	**15**	6	8	5	3	62
21	4	**20**	7	2	28	38
22	**13**	1	13	14	4	54
23	3	3	7	6	**13**	68
24	**18**	4	9	5	5	58
25	2	**8**	2	10	5	73

SMC 2014: Some comments on the pupils' choice of answers as sent to schools in the letter with the results

It is pleasing to see that there has again been an increase in the average mark. This was 53 in 2012, 58 in 2013 and 61 this year. However, we should not be complacent. You can see from the table included with your results how the national distribution of answers to the questions compares with those of your students. We hope you will find the time to look at this information, and, in particular, that you will discuss with your pupils questions where they were not as successful as you might have hoped.

For once, the Problems Group set a Question 1 that almost everyone managed to get correct. This was a question that is accessible to pupils in the first year of high school, but we hope that your A-level pupils found a quick method. There was just one other question which more than three-quarters of the candidates got right. This was Question 8. It is instructive to think about the reason for the good result on this question compared with the rather disappointing outcome on earlier questions.

In Question 8 there is no straightforward way to guess a plausible answer. Once you have chosen your method, you need to spend some time to do a careful sum. It is good to see that almost 90% of the candidates succeeded with this.

In earlier questions, careless reading or not enough thought can easily lead to the wrong answer. In Question 3, for example, 52 is a plausible guess, which would have been correct for any day of the week other than Wednesday. It seems that over a third of the candidates went for this 'obvious' answer without stopping to think about the relevance of the information that December 31st 1997 was a Wednesday. It is more difficult to understand the thought processes of the 10% who thought that there were only 51 Wednesdays in 1997.

In Question 4, it looks as though one-fifth of the candidates multiplied $\frac{1}{4}$ by $\frac{1}{5}$ to obtain $\frac{1}{20}$, and then argued that, if one twentieth of the original sum is £15, then the original sum was £300; the right answer to a different question! If a significant number of your pupils made this mistake, do remind them that even under the time pressure of an exam, it pays to read the questions carefully, and to allow yourself a little time to think.

The SMC marks

The profile of marks obtained is shown below.

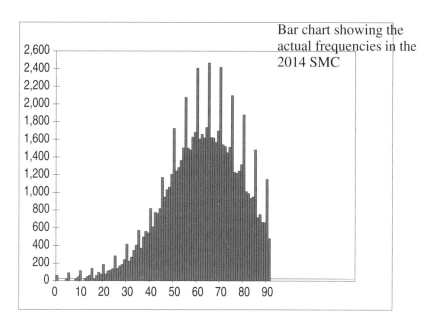

Bar chart showing the actual frequencies in the 2014 SMC

Since 2012, the UKMT has awarded certificates to the top 60% of SMC students. On this basis the cut-off marks were set at

GOLD – 84 or over   SILVER – 71 to 83   BRONZE – 57 to 70

Candidates who scored 101 or more were invited to take part in BMO 1 and those who scored 89 or more were invited to take part in the Senior Kangaroo.

A sample of one of the certificates is shown below.

THE UNITED KINGDOM SENIOR MATHEMATICAL CHALLENGE

The Senior Mathematical Challenge (SMC) is run by the UK Mathematics Trust. The SMC encourages mathematical reasoning, precision of thought, and fluency in using basic mathematical techniques to solve interesting problems. It is aimed at those in full-time education and with sufficient mathematical competence to undertake a post-16 course.

The problems on the SMC are designed to make students think. Most are accessible, yet still challenge those with more experience; they are also meant to be memorable and enjoyable.

Mathematics controls more aspects of the modern world than most people realise—from iPods, cash machines, telecommunications and airline booking systems to production processes in engineering, efficient distribution and stock-holding, investment strategies and 'whispering' jet engines. The scientific and industrial revolutions flowed from the realisation that mathematics was both the language of nature, and also a way of analysing—and hence controlling—our environment. In the last fifty years old and new applications of mathematical ideas have transformed the way we live.

All these developments depend on mathematical thinking—a mode of thought whose essential style is far more permanent than the wave of technological change which it has made possible. The problems on the SMC reflect this style, which pervades all mathematics, by encouraging students to think clearly about challenging problems.

The SMC was established as the National Mathematics Contest in 1961. In recent years there have been over 100,000 entries from around 2000 schools and colleges. Certificates are awarded to the highest scoring 60% of candidates (Gold : Silver : Bronze 1 : 2 : 3).

The UKMT is a registered charity. Please see our website www.ukmt.org.uk for more information. Donations to support our work would be gratefully received; a link for on-line donations is below.

www.donate.ukmt.org.uk

The Next Stages

Subject to certain conditions, candidates who obtained a score of 101 or over in the 2014 Senior Mathematical Challenge were invited to take the British Mathematical Olympiad Round One and those who scored from 89 to 100 were invited to take part in the Senior Kangaroo. It makes use of Kangaroo questions as well as a few others and is not a multiple choice paper but can be marked by character recognition as all the answers are three-digit numbers.

SENIOR 'KANGAROO' MATHEMATICAL CHALLENGE

Friday 28th November 2014

Organised by the United Kingdom Mathematics Trust

The Senior Kangaroo paper allows students in the UK to test themselves on questions set for the best school-aged mathematicians from across Europe and beyond.

RULES AND GUIDELINES (to be read before starting):

1. Do not open the paper until the Invigilator tells you to do so.

2. Time allowed: **1 hour**.

3. The use of rough paper is allowed; **calculators** and measuring instruments are **forbidden**.

4. **Use B or HB pencil only** to complete your personal details and record your answers on the machine-readable Answer Sheet provided. **All answers are written using three digits, from 000 to 999.** For example, if you think the answer to a question is 42, write 042 at the top of the answer grid and then code your answer by putting solid black pencil lines through the 0, the 4 and the 2 beneath.

   Please note that the machine that reads your Answer Sheet will only see the solid black lines through the numbers beneath, not the written digits above. You must ensure that you code your answers or you will not receive any marks. There are further instructions and examples on the Answer Sheet.

5. The paper contains 20 questions. Five marks will be awarded for each correct answer. There is no penalty for giving an incorrect answer.

6. The questions on this paper challenge you **to think**, not to guess. Though you will not lose marks for getting answers wrong, you will undoubtedly get more marks, and more satisfaction, by doing a few questions carefully than by guessing lots of answers.

Enquiries about the Senior Kangaroo should be sent to:
Maths Challenges Office, School of Maths Satellite,
University of Leeds, Leeds, LS2 9JT
Tel. 0113 343 2339
www.ukmt.org.uk

1. Three standard dice are stacked in a tower so that the numbers on each pair of touching faces add to 5. The number on the top of the tower is even. What is the number on the base of the tower?

2. How many prime numbers p have the property that $p^4 + 1$ is also prime?

3. Neil has a combination lock. He knows that the combination is a four-digit number with first digit 2 and fourth digit 8 and that the number is divisible by 9. How many different numbers with that property are there?

4. In the diagram, triangle ABC is isosceles with $CA = CB$ and point D lies on AB with $AD = AC$ and $DB = DC$. What is the size in degrees of angle BCA?

    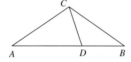

5. Six of the seven numbers 11, 20, 15, 25, 16, 19 and 17 are divided into three groups of two numbers so that the sum of the two numbers in each group is the same. Which number is not used?

6. The numbers x, y and z satisfy the equations $x^2yz^3 = 7^3$ and $xy^2 = 7^9$. What is the value of $\frac{xyz}{7}$?

7. A table of numbers has 21 columns labelled 1, 2, 3, ..., 21 and 33 rows labelled 1, 2, 3, ..., 33. Every element of the table is equal to 2. All the rows whose label is not a multiple of 3 are erased. All the columns whose label is not an even number are erased. What is the sum of the numbers that remain in the table?

8. Andrew wishes to place a number in each circle in the diagram. The sum of the numbers in the circles of any closed loop of length three must be 30. The sum of the numbers in the circles of any closed loop of length four must be 40. He places the number 9 in the circle marked X. What number should he put in the circle marked Y?

    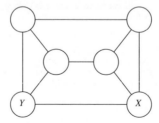

9. Each of the cubes in the diagram has side length 3 cm. The length of AB is $\sqrt{k}$ cm. What is the value of k?

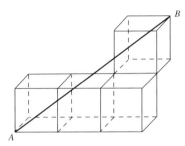

10. A Mathematical Challenge consists of five problems, each of which is worth a different whole number of marks. Carl solved all five problems correctly. He scored 10 marks for the two problems with the lowest numbers of marks and 18 marks for the two problems with the highest numbers of marks. How many marks did he score for all five problems?

11. The mean weight of five children is 45 kg. The mean weight of the lightest three children is 42 kg and the mean weight of the heaviest three children is 49 kg. What is the median weight of the children in kg?

12. On Old MacDonald's farm, the numbers of horses and cows are in the ratio 6:5, the numbers of pigs and sheep are in the ratio 4:3 and the numbers of cows and pigs are in the ratio 2:1. What is the smallest number of animals that can be on the farm?

13. The diagram shows a circle with diameter AB. The coordinates of A are $(-2, 0)$ and the coordinates of B are $(8, 0)$. The circle cuts the y-axis at points D and E. What is the length of DE?

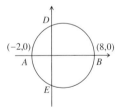

14. Rachel draws 36 kangaroos using three different colours. 25 of the kangaroos are drawn using some grey, 28 are drawn using some pink and 20 are drawn using some brown. Five of the kangaroos are drawn using all three colours. How many kangaroos did she draw that use only one colour?

15. A box contains seven cards numbered from 301 to 307. Graham picks three cards from the box and then Zoe picks two cards from the remainder. Graham looks at his cards and then says "I know that the sum of the numbers on your cards is even". What is the sum of the numbers on Graham's cards?

16. The numbers x, y and z satisfy the equations $x + y + z = 15$ and $\dfrac{1}{x} + \dfrac{1}{y} + \dfrac{1}{z} = 0$. What is the value of $x^2 + y^2 + z^2$?

17. In the diagram, $PQRS$ is a square. M is the midpoint of PQ. The area of the square is k times the area of the shaded region. What is the value of k?

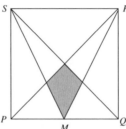

18. Twenty-five workmen have completed a fifth of a project in eight days. Their foreman then decides that the project must be completed in the next 20 days. What is the smallest number of additional workmen required to complete the project on time?

19. In the long multiplication sum shown, each asterisk stands for one digit.

   What is the sum of the digits of the answer?

$$
\begin{array}{r}
***\\
\times\ ***\\
\hline
22**\\
90*0\\
2\\
\hline
56***
\end{array}
$$

20. In the quadrilateral $PQRS$ with $PQ = PS = 25$ cm and $QR = RS = 15$ cm, point T lies on PQ so that $PT = 15$ cm and so that TS is parallel to QR. What is the length in centimetres of TS?

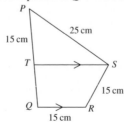

Further remarks

A solutions leaflet was provided.

SENIOR 'KANGAROO' MATHEMATICAL CHALLENGE

Friday 28th November 2014

Organised by the United Kingdom Mathematics Trust

SOLUTIONS

134

1. **5** First note that, on a standard die, the numbers on opposite faces add to 7. Let the number on the top of the tower be n. The numbers on the touching faces going down the tower are then $7 - n$, $5 - (7 - n) = n - 2$, $7 - (n - 2) = 9 - n$ and $5 - (9 - n) = n - 4$ respectively. The bottom number is $7 - (n - 4) = 11 - n$. The numbers on a die are 1 to 6 so $11 - n \leqslant 6$ and hence $n \geqslant 5$. The question states that n is even so $n = 6$. Hence the number on the bottom of the tower is $11 - n = 5$. (It is easy to check that when $n = 6$ all the numbers going down the tower are values that can appear on the face of a standard die.)

2. **1** All prime numbers p greater than 2 are odd. For these numbers, $p^4 + 1$ is even and greater than 2 and so not prime. However, $2^4 + 1 = 17$ which is prime. Hence only one prime number, namely 2, has the desired property.

3. **11** A number is divisible by 9 if and only if the sum of its digits is divisible by 9. Let the second and third digits of the combination be x and y respectively. Hence $10 + x + y$ is divisible by 9. Since $0 \leqslant x \leqslant 9$ and $0 \leqslant y \leqslant 9$ we have $10 + x + y = 18$ or 27. This gives either $x + y = 8$, which has nine different solutions given by $x = 0, x = 1$, and so on up to $x = 8$ or $x + y = 17$ which has two different solutions, namely $x = 8$, $y = 9$ and $x = 9, y = 8$. This means there are $9 + 2 = 11$ different combinations with the desired property.

4. **108** Let $\angle CAB = x°$. Triangle ABC is isosceles with $CA = CB$ so $\angle CBA = x°$. Triangle BCD is also isosceles with $DB = DC$ so $\angle BCD = x°$.

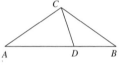

The exterior angle of any triangle is equal to the sum of the interior opposite angles, so $\angle CDA = 2x°$ and hence, since triangle CAD is isosceles, $\angle ACD = 2x°$.

The angle sum of a triangle is 180°, and applying this to triangle CAD we have $x + 2x + 2x = 180$. Therefore $x = 36$ and hence $\angle BCA = 36° + 2 \times 36° = 108°$.

5. **15** Let the number not used be x. The sum of the seven numbers is 123 which is divisible by 3. The six numbers used are divided into three pairs with the same sum so $123 - x$ is also divisible by 3. This means that x is divisible by 3 and the only number in the list that is divisible by 3 is 15. The remaining six numbers can then be paired as 11 and 25, 20 and 16, 19 and 17 all with sum 36.

6. **343** Multiply the two given equations together to obtain $x^2yz^3 \times xy^2 = 7^3 \times 7^9$. Hence $x^3y^3z^3 = 7^{12}$ and so $xyz = 7^4$. Therefore the value of $\dfrac{xyz}{7}$ is $7^3 = 343$.

7. **220** When the unwanted rows and columns are erased, 11 rows and 10 columns remain. The table then contains 11×10 entries, all equal to 2. Hence the sum of the numbers remaining in the table is $11 \times 10 \times 2 = 220$.

8. **11** Let the numbers placed in the empty circles be a, b, c and d as shown and let y be the number placed in the circle marked Y. Recall that the number placed in the circle marked X is 9. The sum of the numbers in a closed loop of length 3 is 30 so $a + b + 9 = 30$ and $c + d + y = 30$. Add these two equations to get $a + b + 9 + c + d + y = 60$. However, the sum of the numbers in a closed loop of length 4 is 40. Thus we also have $a + b + c + d = 40$. This tells us that $9 + y = 20$ and hence that $y = 11$ so Andrew should place number 11 in the circle marked Y.

9. **153** Use the three-dimensional version of Pythagoras' Theorem to get
$AB^2 = (3 \times 3)^2 + (2 \times 3)^2 + (2 \times 3)^2 = 81 + 36 + 36$. Hence $AB^2 = 153$ so $k = 153$.

10. **35** Let the number of marks scored for each question be a, b, c, d and e with
$a < b < c < d < e$. The number of marks scored for the two questions with the lowest
number of marks is 10 and so $a + b = 10$. However, $a < b$ and so $b \geqslant 6$. Similarly
$d + e = 18$ and $d < e$ and hence $d \leqslant 8$. So $6 \leqslant b < c < d \leqslant 8$ and therefore
$b = 6, c = 7$ and $d = 8$. So the total number of marks Carl scored is $10 + 7 + 18 = 35$.

11. **48** The total weight of the five children is $5 \times 45\,\text{kg} = 225\,\text{kg}$. Similarly, the total weight of
the three lightest children is $3 \times 42\,\text{kg} = 126\,\text{kg}$ and the total weight of the three heaviest
children is $3 \times 49\,\text{kg} = 147\,\text{kg}$. Since there are five children, the child with the median
weight is both the third lightest and the third heaviest and so has been included in both of
these groups. Hence the median weight is $126\,\text{kg} + 147\,\text{kg} - 225\,\text{kg} = 48\,\text{kg}$.

12. **123** Since the ratio of the numbers of horses to cows is 6 : 5, the number of cows must be a
multiple of 5. Since the ratio of the numbers of cows to pigs is 2 : 1, the number of pigs
must also be a multiple of 5. Also, since the ratio of the numbers of pigs to sheep is 4 : 3,
the number of pigs is a multiple of 4. Hence the number of pigs is a multiple of 20. The
smallest multiple of 20 is 20 itself and one can check that 20 pigs is feasible, with the
numbers of horses, cows and sheep being 48, 40 and 15 respectively. This gives the
smallest number of animals on the farm as 123.

13. **8** The diameter of the circle is $8 - (-2) = 10$ units so the radius is 5 units. The centre of
the circle is at X, the midpoint of AB, with coordinates $(3, 0)$. Consider triangle OXD
where O is the origin. This is a right-angled triangle with one side 3 units and hypotenuse
5 units so has third side 4 units. Thus the coordinates of D are $(0, 4)$ and the coordinates
of E will be $(0, -4)$ by symmetry. Hence the length of DE is $4 - (-4) = 8$ units.

14. **4** Each kangaroo is drawn using one, two or three colours. So, for example, the number 25
of kangaroos drawn using some grey includes the kangaroos that are only grey, those that
are grey and exactly one other colour and those that are all three colours. Therefore, by
adding 25, 28 and 20 we count those kangaroos with just one colour once, we count those
that have exactly two colours twice and those that have all three colours three times.
Hence $25 + 28 + 20 = 36 + $ (number with exactly two colours) $+ 2 \times$ (number with
three colours) which simplifies to $73 = 36 + $ (number with exactly two colours) $+$
2×5. Hence the number drawn with exactly two colours is 27 and so the number drawn
with only one colour is $36 - 27 - 5 = 4$.

15. **912** To be certain that the sum of the numbers on Zoe's two cards is even, the four cards that she
chose from cannot contain cards of different parity (that is, they are all odd or all even).
The original set of seven cards contained four odd-numbered cards and three even-
numbered cards, so the only way a set of four cards all with the same parity can remain is if
Graham chose the three even-numbered cards. Hence the sum of the numbers on Graham's
cards is $302 + 304 + 306 = 912$.

16. **225** Multiply each term of the second equation by xyz to obtain $yz + xz + xy = 0$. Square each side of the first equation to obtain $(x + y + z)^2 = 15^2$. So $x^2 + 2xy + 2xz + y^2 + 2yz + z^2 = 225$ and hence $x^2 + y^2 + z^2 = 225$.

17. **12** Let the length of the sides of the square be 2 units so its area is 4 units^2. Introduce points X, Y and Z as shown on the diagram where XM and YZ are parallel to SP and let the length of PZ be x units. The triangles PZY, PMX and PQR are all similar and isosceles so $YZ = x$ and $XM = 1$. Also triangles SPM and YZM are similar so $\dfrac{2}{1} = \dfrac{x}{1-x}$ which has solution $x = \frac{2}{3}$. The shaded area is then $2 \times \frac{1}{2} \times 1 \times \left(1 - \frac{2}{3}\right) = \frac{1}{3}$. Hence the area of the square is $4 \div \frac{1}{3} = 12$ times the shaded area and so $k = 12$.

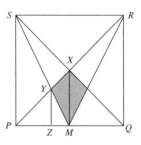

18. **15** The total number of 'man-days' of work required for the project is $5 \times 25 \times 8 = 1000$. The number of 'man-days' completed is $25 \times 8 = 200$ leaving 800 to be completed. To finish this in 20 days requires $800 \div 20 = 40$ workmen and so an extra $40 - 25 = 15$ workmen are required.

19. **16**

```
      ***
 ×    ***
    22**
    90*0
  **2**
  56***
```

The figures 2, 0 and 2 in the hundreds column lines 3, 4 and 5 of the calculation are not large enough to create any carry into the thousands column. Hence the first two missing figures in the third row of working must add with 2 and 9 to give 56 and so are 4 and 5. Note also that the final two digits in that row must be zeros from the structure of the sum so the third row of working is 45200. This means that one of the original 3-digit multiplicands is a 3-digit factor of 452 and so is 452,

226 or 113. The first row of the working is a 4-digit number starting 22 and so is 2260 as it is also a multiple of the same multiplicand. This means that this multiplicand is 452 and that the completed sum is as shown on the right.

Hence the sum of the digits of the answer is $5 + 6 + 5 + 0 + 0 = 16$.

```
      452
 ×    125
     2260
     9040
    45200
    56500
```

20. **24** Draw in line PR as shown and let X be the point where PR intersects TS. The corresponding sides of $\triangle PQR$ and $\triangle PSR$ are equal and so $\triangle PQR$ and $\triangle PSR$ are congruent (SSS). Hence $\angle PRQ = \angle PRS$. Note also that, because TS and QR are parallel, $\angle PRQ = \angle RXS$ since they are alternate angles. This means that $\angle PRS = \angle RXS$ and so $\triangle XRS$ is isosceles and hence $XS = RS = 15$ cm.

TX is parallel to QR and so $\angle PTX = \angle PQR$ and $\angle PXT = \angle PRQ$ using corresponding angles. This means that $\triangle PTX$ and $\triangle PQR$ are similar and so $TX : QR = PT : PQ$ which gives $TX : 15 = 15 : 25$ so $TX = 9$ cm.

Hence $TS = TX + XS = 9$ cm $+ 15$ cm $= 24$ cm.

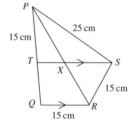

Certificates

These were awarded at two levels, Merit and Participation.

Senior Kangaroo 2014

of

received a

CERTIFICATE of PARTICIPATION

Chairman, United Kingdom Mathematics Trust

THE UKMT SENIOR KANGAROO

The Senior Kangaroo is one of the follow-on rounds for the Senior Mathematical Challenge (SMC) and is organised by the UK Mathematics Trust (UKMT). Around 3,500 high-scoring students in the SMC are invited to participate in the Senior Kangaroo and to test themselves on questions set for the best school-aged mathematicians from across Europe and beyond.

The Senior Kangaroo is a one-hour examination comprising 20 questions; all answers are written using 3 digits, from 000 to 999. The problems involve amusing and thought-provoking situations which require the use of logic as well as mathematical understanding.

The UKMT is a registered charity.
For more information please see our website www.ukmt.org.uk
Donations to support our work would be gratefully
received and can be made by visiting
www.donate.ukmt.org.uk

138

Mathematical Olympiad for Girls

The UK Mathematical Olympiad for Girls (UK MOG) is held annually to identify students to engage in training for European Girls Maths Olympiad (EGMO). Students who are not involved in training are still eligible for selection for the team.

The 2014 MOG paper was held on 23rd September. The time allowed was $2\frac{1}{2}$ hours. The question paper and solutions follow, and a prize-winner list.

United Kingdom Mathematics Trust
UK Mathematical Olympiad for Girls
Tuesday 23rd September 2014

Instructions

1. Do not turn over until told to do so.
2. Time allowed: $2\frac{1}{2}$ hours.
3. Each question carries 10 marks. Full marks will be awarded for written solutions – not just answers – with complete proofs of any assertions you may make.

    Marks awarded will depend on the clarity of your mathematical presentation. Work in rough first, and then write up your best attempt.
4. Partial marks may be awarded for good ideas, so try to hand in everything that documents your thinking on the problem – the more clearly written the better.

    However, one complete solution will gain more credit than several unfinished attempts.
5. Earlier questions tend to be easier. Some questions have two parts. Part (a) introduces results or ideas useful in solving part (b).
6. The use of rulers and compasses is allowed, but calculators and protractors are forbidden.
7. Start each question on a fresh sheet of paper. Write on one side of the paper only.

    On each sheet of working write the number of the question in the top left-hand corner and your name, initials and school in the top right-hand corner.
8. Complete the cover sheet provided and attach it to the front of your script, followed by your solutions in question number order.
9. Staple all the pages neatly together in the top left-hand corner.
10. To accommodate candidates sitting in other time zones, please do not discuss the paper on the internet until 08:00 BST on Wednesday 24th September.

Enquiries about the Mathematical Olympiad for Girls should be sent to:
UKMT, School of Mathematics Satellite, University of Leeds, Leeds LS2 9JT
0113 343 2339 : enquiry@ukmt.org.uk : www.ukmt.org.uk

1. A chord of a circle has length $3n$, where n is a positive integer. The segment cut off by the chord has height n, as shown.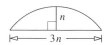

   What is the smallest value of n for which the radius of the circle is also a positive integer?

2. (a) Some strings of three letters have the property that all three letters are the same; for example, LLL is such a string.

   How many strings of three letters do not have all three letters the same?

   (b) Call a number 'hexed' when it has a recurring decimal form in which both the following conditions are true.

   (i) The shortest recurring block has length six.

   (ii) The shortest recurring block starts immediately after the decimal point.

   For example, $987.\dot{1}2345\dot{6}$ is a hexed number (the dots indicating that 123456 is a recurring block).

   How many hexed numbers are there between 0 and 1?

3. A large whiteboard has 2014 + signs and 2015 − signs written on it. You are allowed to delete two of the symbols and replace them according to the following two rules.

   (i) If the two deleted symbols are the same, then replace them by +.

   (ii) If the two deleted symbols are different, then replace them by −.

   You repeat this until there is only one symbol left. Which symbol is it?

4. (a) In the quadrilateral $ABCD$, the sides AB and DC are parallel, and the diagonal BD bisects angle ABC. Let X be the point of intersection of the diagonals AC and BD.

   Prove that $\dfrac{AX}{XC} = \dfrac{AB}{BC}$.

   (b) In triangle PQR, the lengths of all three sides are positive integers. The point M lies on the side QR so that PM is the internal bisector of the angle QPR. Also, $QM = 2$ and $MR = 3$.

   What are the possible lengths of the sides of the triangle PQR?

5. The AM-GM inequality states that, for positive real numbers $x_1, x_2, \ldots, x_n$,

$$\frac{x_1 + x_2 + \ldots + x_n}{n} \geqslant \sqrt[n]{x_1 x_2 \ldots x_n}$$

and equality holds if and only if $x_1 = x_2 = \ldots = x_n$.

   (a) Prove that $\dfrac{a}{b} + \dfrac{b}{c} + \dfrac{c}{a} \geqslant 3$ for all positive real numbers a, b and c, and determine when equality holds.

   (b) Find the minimum value of $\dfrac{a^2}{b} + \dfrac{b}{c^2} + \dfrac{c}{a}$ where a, b and c are positive real numbers.

Time allowed: $2\frac{1}{2}$ hours

Mathematical Olympiad for Girls: Solutions

These are polished solutions and do not illustrate the process of failed ideas and rough work by which candidates may arrive at their own solutions. Some of the solutions include comments, which are intended to clarify the reasoning behind the selection of a particular method.

There is an explanation of the strategy of marking Olympiad papers in the IMOK section of this book.

Enquiries about the Mathematical Olympiad for Girls should be sent to: UKMT, School of Mathematics Satellite, University of Leeds, Leeds LS2 9JT. 0113 343 2339 enquiry@ukmt.org.uk

General Comments

This was the second year of the new format of the Mathematical Olympiad for Girls, in which some questions are split into two parts. The purpose of the first part is to introduce results or ideas needed to answer the second part.

We were pleased that most candidates had a reasonable attempt at all the questions. Many completed part (a) of more than one question. It was also good to see a lot of candidates making a decent attempt to explain and justify their solutions: they grasped that a single numerical answer would not suffice.

We saw many elegant and creative solutions. Many candidates demonstrated real mathematical potential, coming up with their own strategies to solve problems in clearly unfamiliar areas. There were candidates who achieved a high mark for the whole paper and, even more pleasingly, many who produced excellent solutions to individual questions.

One of the most common mistakes that candidates made was trying to argue from special cases, rather than realising that some generality was needed. This was most apparent in Question 3, where a large number of candidates considered only one particular order of deleting the symbols, and in Question 4, where some considered a rectangle or a parallelogram rather than a general trapezium.

The 2014 Mathematical Olympiad for Girls attracted 1630 entries. The scripts were marked on 4th and 5th October in Cambridge by a team of Ben Barrett, Andrew Carlotti, Lax Chan, Andrea Chlebikova, Philip Coggins, Tim Cross, Susan Cubbon, Paul Fannon, James Gazet, Adam P. Goucher, Jo Harbour, Maria Holdcroft, Ina Hughes, Freddie Illingworth, Magdalena Jasicova, Vesna Kadelburg, Lizzie Kimber, David Mestel, Joseph Myers, Vicky Neale, Peter Neumann, Sylvia Neumann, Craig Newbold, Martin Orr, Preeyan Parmar, David Phillips, Linden Ralph, Jenni Sambrook, Eloise Thuey, Jerome Watson and Brian Wilson.

1. A chord of a circle has length $3n$, where n is a positive integer. The segment cut off by the chord has height n, as shown.

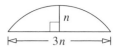

What is the smallest value of n for which the radius of the circle is also a positive integer?

Commentary: The question mentions the radius of the circle, so we should give it a name (such as r) in order to talk about it, and we should find it on the diagram. In fact, it does not conveniently appear on the diagram given in the question, but if we draw the whole circle and its centre then we can find various interesting radii with length r.

It is important to remember that we are told that n and r are positive integers. Without this, it will not be possible to complete the problem.

Solution

Let the centre and radius of the circle be O and r, and let A, B and M be the points marked in the diagram alongside.

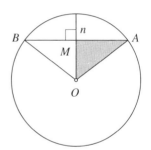

Then $OA = OB = r$ and $OM = r - n$. Since the radius perpendicular to a chord bisects the chord, $AM = \frac{3}{2}n$. Now triangle OAM is right-angled, so we can use Pythagoras' Theorem to give

$$r^2 = \left(\frac{3n}{2}\right)^2 + (r - n)^2$$

$$= \frac{9n^2}{4} + r^2 - 2rn + n^2$$

and hence

$$8rn = 13n^2.$$

Since n is positive and so non-zero, we can divide both sides by n to get

$$8r = 13n,$$

so that

$$r = \frac{13n}{8}.$$

For both r and n to be integers, n needs to be divisible by 8. But n is positive, so the smallest possible value of n is 8 (giving $r = 13$).

142

We saw many excellent solutions to this problem; they all used the crucial idea of introducing the radius. They then went on to find some connection between r and n, perhaps by finding a useful triangle or using a circle theorem.

It is important to be clear about why 8 is the smallest possible value. We wrote r as $\dfrac{13n}{8}$ and this makes it clear that n is a multiple of 8 (strictly speaking because the highest common factor of 8 and 13 is 1, although we did not require candidates to say this explicitly). Some candidates found (true) expressions such as $8r = 13n$ or $r = 1.625n$, but then it is not quite so obvious that n is a multiple of 8, so we needed candidates to explain how they knew that 8 is the *smallest* possible value, not just *a* possible value.

---

2. (a) Some strings of three letters have the property that all three letters are the same; for example, LLL is such a string.

   How many strings of three letters do not have all three letters the same?

   (b) Call a number 'hexed' when it has a recurring decimal form in which both the following conditions are true.

   (i) The shortest recurring block has length six.

   (ii) The shortest recurring block starts immediately after the decimal point.

   For example, $987.\dot{1}2345\dot{6}$ is a hexed number (the dots indicating that 123456 is a recurring block).

   How many hexed numbers are there between 0 and 1?

Commentary:

(a) We are asked to find the number of strings which do not have a certain property. One way to do this is to count how many strings do have the property, and then subtract this from the total number of strings.

It is also possible to count the number of strings with the given property directly. One way to do this is to think about possible starting combinations of letters: how many strings start with AA and how many with AB? Another way is to identify different possible types of strings – these are *xyz*, *xxy*, *xyx* and *yxx*, where *x*, *y* and *z* are different letters – and count each type separately. It helps to realise that the last three are going to give the same number. The last two approaches become considerably less practical when we are considering longer strings, such as in part (b). In general, excluding unwanted objects is often easier than counting the ones we want.

(b) We need to think about what a hexed number could look like. After the decimal point, it has six digits which then repeat. Since we are only looking for numbers between 0 and 1, the number before the decimal point is 0. So it looks like 0.*abcdef abcdef*... .

However, there are restrictions on what the recurring block *abcdef* can be. The requirement that this should be the shortest recurring block means that it cannot have a shorter repeating block within it. For example, 0.12121212 ... is not a hexed number because although the block '121212' repeats, the shorter block '12' also repeats. However, 0.121233 121233 ... is a hexed number.

Now, counting all hexed numbers is the same as counting all possible blocks of length six which do not contain a shorter repeated sub-block. We can do this by identifying all the blocks that need to be excluded and taking this away from the total number of possible blocks of length six.

We need to exclude the following types of blocks:

(i)   those where all six digits are the same;

(ii) those where a sub-block of two *different* digits repeats, such as *ababab* (blocks with all digits the same have already been excluded in (i));

(iii) those where a sub-block of three digits repeats, but not those where all three digits are the same (because these have already been excluded in (i)). Note that this is basically what we counted in part (a).

It is not possible to have sub-blocks of length 4 or 5, as 4 and 5 are not factors of 6.

The question does not specify whether 'between 0 and 1' includes the endpoints. But, 0.$\dot{0}$ and 0.$\dot{9}$ are not hexed numbers, so they are going to be excluded anyway.

Solution to part (a):   The total number of strings of three letters is 26^3, because there are 26 options for each letter. The number of strings where all three letters are the same is 26. So the number of strings which do not have all three letters the same is $26^3 - 26 = 17\,550$.

Solution to part (b):

A hexed number between 0 and 1 is of the form 0.$\dot{a}bcde\dot{f}$, where the block *abcdef* contains no shorter repeating sub-block. Hence counting hexed numbers between 0 and 1 is the same as counting blocks of length six, made up of digits 0 to 9, which do not contain any repeated sub-blocks.

The total number of blocks of length six is 10^6, because there are 10 options for each digit. We need to exclude those blocks that have repeating sub-blocks of length one, two or three.

(i)   If there is a repeating sub-block of length one, this means that all six digits are the same. There are 10 such blocks.

144

(ii) If there is a repeating sub-block of length two, the whole block looks like *ababab*, where *a* and *b* are different digits. There are 10 options for *a* and for each of them there are 9 options for *b*. So the number of blocks of the form *ababab*, where *a* and *b* are different, is $10 \times 9 = 90$.

(iii) The repeating sub-blocks of length three are precisely all strings of three digits except those where all three digits are the same. We essentially counted this in part (a) (although there we had 26 letters, rather than the 10 digits we have here), so the answer is $10^3 - 10 = 990$.

The total number of hexed numbers is therefore

$$10^6 - 10 - 90 - 990 = 998\,910.$$

Markers' comments

This was the most popular question, with almost half the candidates making substantial progress in part (a) and many attempting part (b) as well.

We saw several sensible approaches to calculating the total number of three-letter strings. Many candidates seemed familiar with the 'multiplication rule' – counting the number of options for each letter and then multiplying those together – although some obtained 3^{26} instead of 26^3. Another common approach was to count the number of possibilities when the first letter is A, and then multiply the answer by 26. A common mistake here is to forget that the letters do not need to appear in alphabetical order; so having listed AAA, AAB, ..., and then having moved onto the strings starting with AB, some candidates forgot to include ABA before ABB.

One advantage of counting all strings of three digits and subtracting the ones we don't want is that it easily extends to counting all possible strings of six digits, which we need in part (b).

The phrase 'the shortest recurring block has length six' seemed to confuse some candidates, who interpreted it as meaning that a hexed number can have recurring blocks of any length greater than or equal to six. What it means is that there is a recurring block of length six, and no shorter recurring block.

We were happy to accept correct expressions instead of numerical values, so answers in the form $26^3 - 26$ or $25 \times 26 \times 27$ received full marks (even where candidates had attempted the calculation and made a slip in the arithmetic). We also accepted answers from candidates who used alphabets with different numbers of letters (not necessarily 26), or thought that there were nine digits rather than ten, as long as the answer was correct for their preferred choice.

3. A large whiteboard has 2014 + signs and 2015 − signs written on it. You are allowed to delete two of the symbols and replace them according to the following two rules.

  (i) If the two deleted symbols are the same, then replace them by +.

  (ii) If the two deleted symbols are different, then replace them by −.

  You repeat this until there is only one symbol left. Which symbol is it?

Commentary:

It may be useful to start investigating what happens by looking at a particular order of deleting the symbols. For example, we could delete 1007 pairs of − signs, leaving only one − sign and 3021 + signs (because 2014 + 1007 = 3021). Then keep deleting pairs of + signs until there is only one left. This leaves one + and one − sign, and deleting those leaves you with a − sign. There are several other examples leading to the same answer.

Another useful approach may be to investigate what happens if you start with fewer symbols, but we should think carefully about the significance of the specific numbers 2014 and 2015.

At this point we might think that we have solved the problem. However, there is a very common trap here! The description above shows that we *can* end up with one − sign. However it does not guarantee that this will happen *regardless of the order in which you delete the symbols*. To solve the problem, we must show either that we always end up with one − sign, or that the final symbol depends on the order in which we deleted them.

We need an argument that does not rely on a specific example. The answer probably has something to do with the fact that there is an odd number of − signs, which seems to make it impossible to eliminate the final one. So one approach is to look at whether, after each step, the number of each type of symbol is even or odd (this is called the *parity* of the number).

Alternatively, the rules of the game should remind us of the rules of multiplication of positive and negative numbers. We present two solutions, one using each of these ideas.

Solution

Method 1

At each step, one of the following three things happens:

(i)   two + signs are replaced by one + sign;

(ii)   two − signs are replaced by one + sign;

(iii) one + sign and one − sign are replaced by one − sign.

In the first and the third case the number of − signs remains the same, while in the second case it decreases by two. Since at the beginning the number of signs was odd, it will remain odd after each step. Hence we can never eliminate all the - signs, so this is the last remaining symbol.

Method 2

If we replace every + sign by the number 1 and every − sign with the number −1, then the rules are equivalent to replacing a pair of numbers by their product. The final number remaining is the product of all the numbers we started with, which is $1^{2014} - (-1)^{2015} = -1$. Hence the last remaining symbol is a − sign.

Markers' comments

This was the second most popular question and more than half the candidates scored marks on it.

A large number of candidates did not realise the need for generality and may be disappointed with the number of marks they received. However, we did see many elegantly presented solutions, and many more attempts to explain observed patterns. Some noticed the constant parity of the number of minus signs without realising that this essentially solves the problem. Many thought that the key issue is that the number of minus signs is one more than the number of plus signs; this is not the case.

Another important point to note here is the need for clear presentation and commentary. A table containing the numbers of plus and minus signs at various stages can be difficult to interpret. It is much better to describe the strategy, such as saying which of the three rules is being used at each stage.

---

4. (a) In the quadrilateral *ABCD*, the sides *AB* and *DC* are parallel, and the diagonal *BD* bisects angle *ABC*. Let *X* be the point of intersection of the diagonals *AC* and *BD*.

   Prove that $\dfrac{AX}{XC} = \dfrac{AB}{BC}$.

   (b) In triangle *PQR*, the lengths of all three sides are positive integers. The point *M* lies on the side *QR* so that *PM* is the internal bisector of the angle *QPR*. Also, *QM* = 2 and *MR* = 3.

   What are the possible lengths of the sides of the triangle *PQR*?

Commentary:

(a) The first thing to do when facing a geometry problem is to draw and label a clear diagram. The purpose of a diagram is to convey information, so be sure to draw diagrams clearly, don't make them too small, and label them carefully. It was pleasing to see that most candidates did this as their first step.

It is important not to make additional assumptions that are not explicitly stated in the question. For example, we are told that *AB* and *DC* are parallel, but nothing about *BC* and *AD*; you should therefore draw a trapezium and not a rectangle or a parallelogram.

You should note the convention for labelling quadrilaterals: *ABCD* means that the vertices should appear in that order when going around the shape.

The fact that you are being asked to prove something about ratios of sides should make you think about similar triangles. To find two similar triangles, we need to look at angles. We are told about the angle bisector, and we also have some parallel lines, so there are some good starting points in looking for equal angles. Once we have identified some equal angles, two useful questions to ask are:

(i)   are there any similar triangles?

(ii) are there any isosceles triangles?

There are several alternative ways to prove this result, for example by using the sine rule or areas of triangles *ABX* and *BXC*. This does not use point *D* at all. In fact, the result in part (a) is really about triangle *ABC* and it is called the *Angle Bisector Theorem*. It tells us something about the ratio in which an angle bisector divides the opposite side.

(b)   The set-up looks similar to part (a), in that there is an angle bisector and the point at which it intersects the opposite side of the triangle. So you should be looking to use the result from part (a). If you cannot immediately see how, you can add another point to the diagram so that it looks like the one from part (a). To do this, you need to extend the line *PM* to a point *D* such that *DQ* is parallel to *RP*.

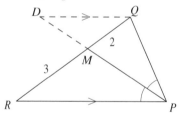

You should now be able to see that part (a) applies, so that

$$\frac{3}{2} = \frac{RP}{PQ}.$$

We now have the ratio of the lengths of two sides of the triangle. Remembering that they need to be integers, you can deduce that *PQ* has to be an even number and *RP* a multiple of 3. But this still leaves infinitely many possibilities. Is there any property of a triangle that means that only some of these are actually possible?

In any triangle, the sum of the lengths of any two sides is greater than the length of the third side; this is called the *triangle inequality*. Combining it with the ratio we found above leads to a finite number of possible combinations of sides.

Solution to part (a)

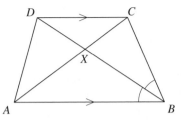

Since AB is parallel to DC, $\angle ABD = \angle CDB$ (alternate angles). In triangles ABX and CDX, the angles at X are also equal (vertically opposite angles). Hence the two triangles are similar, and

$$\frac{AX}{CX} = \frac{AB}{CD}.$$

We are also given that $\angle ABD = \angle DBC$, so that $\angle CDB = \angle DBC$. This implies that $BC = CD$ (sides opposite equal angles in triangle BCD). Combining this with the above, we get

$$\frac{AX}{XC} = \frac{AB}{BC},$$

as required.

Solution to part (b)

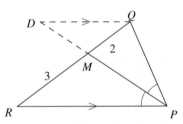

Using the result from part (a) (as shown in the diagram), we get

$$\frac{RP}{PQ} = \frac{3}{2}.$$

But PQ, RP and $QR(= 5)$ are sides of a triangle, so they satisfy the three triangle inequalities

$$PQ + RP > 5,$$
$$PQ + 5 > RP,$$
$$\text{and} \quad RP + 5 > PQ.$$

Substituting $RP = \frac{3}{2}PQ$, we get

$$\tfrac{5}{2}PQ > 5$$

$$PQ + 5 > \tfrac{3}{2}PQ,$$

$$\text{and} \quad \tfrac{3}{2}PQ + 5 > PQ.$$

The third inequality is always true, but from the first two inequalities we obtain $PQ > 2$ and $PQ < 10$. Since PQ is an integer, the possible values are 3, 4, 5, 6, 7, 8 and 9. However, $RP = \tfrac{3}{2}PQ$ is also an integer, so PQ is even. Therefore there are only three possibilities:

$$PQ = 4, \, RP = 6 \text{ and } QR = 5;$$

$$PQ = 6, \, RP = 9 \text{ and } QR = 5;$$

$$PQ = 8, \, RP = 12 \text{ and } QR = 5.$$

Markers' comments

Given that students are often intimidated by geometry, it was great to see a large number of attempts at this question. Many candidates did realise the need for some generality and tried to avoid drawing squares and rectangles. Unfortunately we saw other, more subtle incorrect assumptions, such as: just because the diagonal BD bisects the angle at B, it also bisects the angle at D; or that the other diagonal will bisect angle A; or that the diagonals bisect each other. It is a useful exercise to try drawing some accurate diagrams to explore whether any of those are necessarily true; dynamic geometry software can be a very useful tool for further exploration.

This question revealed some common misconceptions about geometrical notation. For example, a number of candidates wrote statements like

$$\frac{AX}{BX} = \frac{A}{B},$$

suggesting that they thought of the letters used to label points as variables, so that AX represents the product of A and X. We saw similar mistakes when dealing with angles, such as '$\angle ABD = \angle DBC$ so $\angle A = \angle C$' ('cancelling' B and D). Since most senior Olympiad geometry problems do not come with a diagram, it is important that candidates can interpret geometrical notation and terminology correctly.

Some candidates were clearly familiar with the angle bisector theorem and simply quoted it in part (a), without giving a proof. As the intention of part (a) is to introduce potentially new concepts which can help in part (b), this was not penalised, as long as they clearly stated which triangle the theorem was being applied to.

In part (b) a fair number of candidates seemed familiar with the triangle inequality, but some did not realise that two of the inequalities were needed to limit the number of solutions. One common pitfall was to use the triangle inequality to *check* that $PQ = 4$, $PQ = 6$ and $PQ = 8$ work, but $PQ = 10$ does not, and conclude that all solutions have been found. This was penalised heavily, since this method does not *prove* that no larger value of PQ can work.

Having found the three possible pairs of sides, we should really check that they all satisfy the conditions of the question, that is, in each of the resulting triangles the angle bisector does indeed divide the side QR in the ratio $2 : 3$. This is easily done by using the converse of the angle bisector theorem but, since this was not included in the proof of the theorem in part (a), it was not required for full marks on this occasion.

---

5. The AM-GM inequality states that, for positive real numbers $x_1, x_2, \ldots, x_n$,

$$\frac{x_1 + x_2 + \ldots + x_n}{n} \geqslant \sqrt[n]{x_1 x_2 \ldots x_n}$$

and equality holds if and only if $x_1 = x_2 = \ldots = x_n$.

(a) Prove that

$$\frac{a}{b} + \frac{b}{c} + \frac{c}{a} \geqslant 3$$

for all positive real numbers a, b and c, and determine when equality holds.

(b) Find the minimum value of

$$\frac{a^2}{b} + \frac{b}{c^2} + \frac{c}{a}$$

where a, b and c are positive real numbers.

Commentary:

(a) The statement of the AM-GM inequality looks rather intimidating. In part (a) there seem to be three numbers added together, so let's see what the inequality says when $n = 3$. If we call the three numbers x, y and z, then

$$\frac{x + y + z}{3} \geqslant \sqrt[3]{xyz}.$$

where x, y and z are positive real numbers. It seems useful to set

$$x = \frac{a}{b}, y = \frac{b}{c}, z = \frac{c}{a}.$$

AM-GM also tells us that equality holds if, and only if, $x = y = z$. (It is quite straightforward to check that if $x = y = z$ then both sides are x and so we have equality. What is true and useful, but much less obvious, is that this is the only way we get equality.)

(b) First of all, we need to understand what the question is asking us to do. We want to find the minimum value of the given expression, and this involves doing two things:

(i) finding a value that the expression is always greater than or equal to;

(ii) proving that this value is actually achieved for some a, b and c.

To see why step (ii) is required, think about these examples: x^2 is always greater than -1, but its minimum value is in fact 0; and for a positive number x, $\frac{1}{x}$ is always greater than 0, but it does not have a minimum value at all.

In part (a), we used the AM-GM inequality to show that a similar expression was always greater than or equal to 3, so it seems sensible to start with the same approach, applying AM-GM to $\frac{a^2}{b}$, $\frac{b}{c^2}$ and $\frac{c}{a}$. This gives

$$\frac{1}{3}\left(\frac{a^2}{b} + \frac{b}{c^2} + \frac{c}{a}\right) \geqslant \sqrt[3]{\frac{a^2}{b} \times \frac{b}{c^2} \times \frac{c}{a}}$$

$$= \sqrt[3]{\frac{a}{c}}.$$

Unfortunately, this last expression is not a constant, so it does not give us the minimum value.

It is tempting to hope that we can minimise the left-hand side by finding when equality occurs; it turns out that this is when $c = a$ and $b = a^2$, and then the right-hand side is 1. Unfortunately it might turn out that when $\sqrt[3]{\frac{a}{c}} < 1$ the left-hand side could be smaller than 1 while still being larger than the right-hand side.

The problem seems to be that, this time, the as and cs do not all cancel. Is there a way we can rewrite the expression to make this happen?

Since there is an a^2 in the numerator of the first fraction, we would like two fractions with denominator a. We can achieve this by splitting the last fraction into two:

$$\frac{c}{a} = \frac{c}{2a} + \frac{c}{2a}.$$

This also gives us two cs in the numerator, to cancel the c^2 in the denominator of the second fraction. So we have written the expression as a sum of four fractions, in a way that allows us to apply the AM-GM inequality, as we did in part (a) (but to four quantities this time, rather than three).

There is an alternative approach: to apply the AM-GM inequality to the first two terms, and then again to the result and the third term. We present this as an alternative solution below.

Solution to part (a)

Using the AM-GM inequality with $n = 3$ and setting $x_1 = \frac{a}{b}$, $x_2 = \frac{b}{c}$ and $x_3 = \frac{c}{a}$, we get

$$\frac{1}{3}\left(\frac{a}{b} + \frac{b}{c} + \frac{c}{a}\right) \geqslant \sqrt[3]{\frac{a}{b} \times \frac{b}{c} \times \frac{c}{a}}.$$

The three fractions under the cube root multiply to 1, so this is equivalent to

$$\frac{a}{b} + \frac{b}{c} + \frac{c}{a} \geqslant 3.$$

Equality holds if and only if

$$\frac{a}{b} = \frac{b}{c} = \frac{c}{a}.$$

Multiplying both sides of the first equation by b^2c, we get $abc = b^3$, and multiplying both sides of the second equation by ac^2, we get $abc = c^3$. Therefore $b^3 = c^3$, so that $b = c$, and thus $a = b$, from the first equation. Hence equality holds if and only if $a = b = c$.

Solution to part (b)

Method 1

Rewrite the expression as

$$\frac{a^2}{b} + \frac{b}{c^2} + \frac{c}{2a} + \frac{c}{2a}.$$

Then, using the AM-GM inequality with $n = 4$, we get

$$\frac{a^2}{b} + \frac{b}{c^2} + \frac{c}{2a} + \frac{c}{2a} \geqslant 4\sqrt[4]{\frac{a^2}{b} \times \frac{b}{c^2} \times \frac{c}{2a} \times \frac{c}{2a}}$$

$$= 4\sqrt[4]{\tfrac{1}{4}}$$

$$= \frac{4}{\sqrt{2}}$$

$$= 2\sqrt{2}.$$

To show that $2\sqrt{2}$ is in fact the minimum value, we need to find at least one set of values of a, b and c for which the expression is equal to $2\sqrt{2}$. In the AM-GM inequality, equality holds if and only if all four numbers are equal, so we need

$$\frac{a^2}{b} = \frac{b}{c^2} = \frac{c}{2a}.$$

These are two equations in three variables, but we do not need to solve them – we are just trying to find values of a, b and c that work. So let us put $a = 1$; the equations then become

$$\frac{1}{b} = \frac{b}{c^2} = \frac{c}{2}.$$

Multiplying both sides of the first equation by bc^2, we get $c^2 = b^2$, so that $c = b$, since both b and c are positive. Now from the second equation we obtain $b = c = \sqrt{2}$.

We can check that these values do indeed work:

$$\frac{a^2}{b} + \frac{b}{c^2} + \frac{c}{a} = \frac{1}{\sqrt{2}} + \frac{\sqrt{2}}{2} + \frac{\sqrt{2}}{1}$$

$$= \frac{\sqrt{2}}{2} + \frac{\sqrt{2}}{2} + \sqrt{2}$$

$$= 2\sqrt{2}.$$

Hence the minimum value of the required expression is $2\sqrt{2}$.

Method 2

Applying the AM-GM inequality (with $n = 2$) to $\dfrac{a^2}{b}$ and $\dfrac{b}{c^2}$ gives

$$\frac{a^2}{b} + \frac{b}{c^2} \geqslant 2\sqrt{\frac{a^2}{b} \times \frac{b}{c^2}}$$

$$= \frac{2a}{c}.$$

Applying AM-GM again (also with $n = 2$), this time to $\dfrac{2a}{c}$ and $\dfrac{c}{a}$, we get:

$$\frac{a^2}{b} + \frac{b}{c^2} + \frac{c}{a} \geqslant \frac{2a}{c} + \frac{c}{a}$$

$$\geqslant 2\sqrt{\frac{2a}{c} \times \frac{c}{a}}$$

$$= 2\sqrt{2}.$$

The minimum value is reached if and only if

$$\frac{a^2}{b} = \frac{b}{c^2} \text{ and } \frac{2a}{c} = \frac{c}{a}.$$

Setting $a = 1$, we obtain the same values of b and c from these equations that we found in the first method.

Remark

There are, of course, other values of a, b and c that also give the minimum value of the expression. We can solve the above equations to express b and c in terms of a: $c = \sqrt{2}a$ and $b = \sqrt{2}a^2$.

Markers' comments

It is not surprising that nearly half the candidates did not attempt this question at all, but we saw a good number of solutions to part (a). Only a handful of candidates made progress in part (b).

Several candidates did not realise that n stands for the number of variables involved, rather than the denominator of the fraction in general; they tried writing the left-hand side as $\dfrac{ab + bc + ca}{abc}$ and setting $n = abc$, which is incorrect.

Many candidates tried various values of a, b and c to see what happens. While this is useful in trying to understand the problem, it often led to statements like 'the minimum value is when a, b and c are the smallest possible, so $a = b = c = 1$'. But the question states that a, b, c are positive *real* numbers, so it is impossible to state the smallest possible values of a, b and c. A more subtle, but important, point is that the expression

$$\frac{a}{b} + \frac{b}{c} + \frac{c}{a}$$

is not necessarily made smaller by decreasing a, b or c. For example, when $a = b = c = 2$ the value of the expression is 3, but if we then decrease a to 1 the value increases to $3\frac{1}{2}$.

In part (a), many candidates stated (correctly) that equality holds if $a = b = c$, but either did not know or did not mention that this is the only case when equality occurs. This can be deduced from AM-GM.

Interested students may wish to find out about other similar inequalities. AM-GM refers to the arithmetic mean and the geometric mean of a set of numbers. The arithmetic mean is what we often refer to as 'the mean': the sum of n numbers divided by n. The geometric mean is the nth root of their product. One interesting special case of AM-GM says that if x is a positive real number, then

$$x + \frac{1}{x} \geqslant 2.$$

Other related inequalities involve the quadratic mean (used in statistics) and the harmonic mean.

The Mathematical Olympiad for Girls Prize Winners

The following contestants were awarded prizes:

Olivia Aaronson	St Paul's Girls' School, London
Hannah Black	Stephen Perse Foundation
Rosie Cates	The Perse School
Sophie Crane	Pate's Grammar School
Yanni Du	Stephen Perse Foundation
Katherine Horton	All Hallows School
Olivia Hu	Cardiff Sixth Form College
Claire Jang	Badminton School
Liberty Jones	Monk's Walk School
Sophie Jones	Ibstock Place School
Julie Jouas Yosano	Marymount International School
Tomoka Kan	Westminster School
Kirsten Land	King's College London Maths School
Bridget Langford	The Tiffin Girls' School
Jackie Li	St Paul's Girls' School
Sophie Maclean	Watford Grammar School for Girls
Georgina Majury	Down High School
Eve Pound	King Edward VII School
Sophie Sadler	School of St Helen and St Katharine
Xinyu Shen	Winterbourne International Academy
Polina Skliarevitch	Sidcot School
Marguerite Tong	Westminster School
Alice Vaughan-Williams	Nailsea School
Naomi Wei	City of London School for Girls
Emily Wolfenden	Sir Isaac Newton Sixth Form
Joanna Yass	North London Collegiate
Danshu Zhang	Eltham College
Duo Zhao	Simon Langton Girls' School

British Mathematical Olympiads

Within the UKMT, the British Mathematical Olympiad Subtrust has control of the papers and everything pertaining to them. The BMOS produces an annual account of its events which, for 2014-2015, was edited by James Aaronson (of Trinity College, Cambridge) and Tim Hennock. Much of this report is included in the following pages.

 United Kingdom Mathematics Trust

British Mathematical Olympiad

Round 1 : Friday, 28 November 2014

Time allowed *Three and a half hours.*

Instructions • *Full written solutions – not just answers – are required, with complete proofs of any assertions you may make. Marks awarded will depend on the clarity of your mathematical presentation. Work in rough first, and then write up your best attempt.*

Do not hand in rough work.

• *One* **complete** *solution will gain more credit than several unfinished attempts. It is more important to complete a small number of questions than to try all the problems.*

• *Each question carries 10 marks. However, earlier questions tend to be easier. In general you are advised to concentrate on these problems first.*

• *The use of rulers, set squares and compasses is allowed, but calculators and protractors are forbidden.*

• *Start each question on a fresh sheet of paper. Write on one side of the paper only. On each sheet of working write the number of the question in the top* **left**-*hand corner and your name, initials and school in the top* **right**-*hand corner.*

• *Complete the cover sheet provided and attach it to the front of your script, followed by your solutions in question number order.*

• *Staple all the pages neatly together in the top* **left**- *hand corner.*

• *To accommodate candidates sitting in other time zones, please do not discuss the paper on the internet until 8 am GMT on Saturday 29 November.*

Do not turn over until told to do so.

United Kingdom Mathematics Trust

2014/15 British Mathematical Olympiad
Round 1: Friday, 28 November 2014

1. Place the following numbers in increasing order of size, and justify your reasoning:

$$3^{3^4}, \ 3^{4^3}, \ 3^{4^4}, \ 4^{3^3} \text{ and } 4^{3^4}.$$

Note that a^{b^c} means $a^{(b^c)}$.

2. Positive integers p, a and b satisfy the equation $p^2 + a^2 = b^2$. Prove that if p is a prime greater than 3, then a is a multiple of 12 and $2(p + a + 1)$ is a perfect square.

3. A hotel has ten rooms along each side of a corridor. An olympiad team leader wishes to book seven rooms on the corridor so that no two reserved rooms on the same side of the corridor are adjacent. In how many ways can this be done?

4. Let x be a real number such that $t = x + x^{-1}$ is an integer greater than 2. Prove that $t_n = x^n + x^{-n}$ is an integer for all positive integers n. Determine the values of n for which t divides t_n.

5. Let $ABCD$ be a cyclic quadrilateral. Let F be the midpoint of the arc AB of its circumcircle that does not contain C or D. Let the lines DF and AC meet at P and the lines CF and BD meet at Q. Prove that the lines PQ and AB are parallel.

6. Determine all functions $f(n)$ from the positive integers to the positive integers which satisfy the following condition: whenever a, b and c are positive integers such that $\dfrac{1}{a} + \dfrac{1}{b} = \dfrac{1}{c}$, then

$$\frac{1}{f(a)} + \frac{1}{f(b)} = \frac{1}{f(c)}.$$

The British Mathematical Olympiad 2014-2015

The Round 1 paper was marked by volunteers in December. Below is a list of the prize winners.

Round 1 Prize Winners

The following contestants were awarded prizes:

Gold Medals

Joe Benton	St Paul's School, Barnes, London
Alex Harris	The Perse School, Cambridge
Lawrence Hollom	Churcher's College, Hampshire
Liam Hughes	Robert Smyth Academy, Market Harborough
Samuel Kittle	Simon Langton Boys' G. Sch, Canterbury
Xiangjia Kong	Ermysted's Grammar School, N. Yorks
Kyung Chan Lee	Garden International School, Malaysia
Jiani Li	Ruthin School, Denbighshire
Warren Li	Eton College, Windsor
Milton Lin	Taipei European School
Xiao Ma	Ruthin School, Denbighshire
Bhavik Mehta	Queen Elizabeth's School, Barnet
Harry Metrebian	Winchester College
Yuyang Miao	Ruthin School, Denbighshire
Neel Nanda	Latymer School, London
Philip Peters	Haberdashers' Aske's School for Boys, Herts
Harvey Yau	Ysgol Dyffryn Taf, Carmarthenshire

Silver medals:

Hugo Aaronson	St Paul's School, Barnes, London
Olivia Aaronson	St Paul's Girls' School, Hammersmith
Joseph Boorman	Eton College, Windsor
Jacob Coxon	Magdalen College School, Oxford
Yuan Gao	Anglo-Chinese School, Singapore
Joshua Garfinkel	Latymer School, London
Valeriia Gladkova	St Swithun's School, Winchester
Alex Gunasekera	Magdalen College School, Oxford
Shinichi Hirata	King's College School, Wimbledon
Charlie Hu	City of London School
Jongheon Jeon	Winchester College
Gareth Jones	Clifton College, Bristol
Tomoka Kan	Westminster School

Andrew Kenyon-Roberts	Aberdeen Grammar School
Ricky Li	Fulford School, York
Chris Liu	Winchester College
Jianzhi Long	RDFZ, Beijing
Chen Lu	Eton College, Windsor
Nico Marrin	Wellington College, Berkshire
Conor Murphy	Eltham College, London
Michael Ng	Aylesbury Grammar School
Chikashi Rison	The Leys School, Cambridge
Marcus Roberts	The Grammar School at Leeds
Zhao Runcong	Blundell's School, Devon
Askhat Sarkeev	Caterham School, Surrey
Hoseong Seo	The Perse School, Cambridge
Yukuan Tao	Hampton School, Middlesex
Dang Tu	Bellerbys College Cambridge
Dongxin Wang	Ruthin School, Denbighshire
Martin Ying	Ningbo Xiaoshi High School, China
Joon Young Yoon	North London Collegiate S. Jeju, South Korea
Danshu Zhang	Eltham College, London
Ebony Zhang	The Perse School, Cambridge
Tingjun Zhang	Ashford School, Kent
Yilin Zheng	Ruthin School, Denbighshire
Yuchen Zhu	Ysgol David Hughes, Anglesey

Bronze medals:

Jamie Bamber	The Perse School, Cambridge
Agnijo Banerjee	Grove Academy, Dundee
Jamie Bell	King Edward VI Five Ways School, Birmingham
Jakub Bojdol	St Paul's School, Barnes, London
Erli Cai	Whitgift School, Surrey
Jiaxuan Chen	Wuxi Number 1 High School, China
Cyrus Cheng	Wells Cathedral School, Somerset
Euijin Choi	Wilson's School, Surrey
Wesley Chow	Queen Elizabeth Sixth Form College, Darlington
Mark Cooper	Horsforth School, Leeds
Irene Cortinovis	Churston Ferrers Grammar School, Devon
Sophie Crane	Pate's Grammar School, Cheltenham
Joe Davies	Lawrence Sheriff School, Rugby
Edwin Fennell	Hills Road VI Form College, Cambridge
James Fraser	Winchester College

Harry Goodburn	Wilson's School, Surrey
Luke Gostelow	Hampton School, Middlesex
Kiwan Hyun	North London Collegiate Sch. Jeju, South Korea
Stephen Jones	Magdalen College School, Oxford
Balaji Krishna	Stanwell School, Vale of Glamorgan
Kirsten Land	King's College London Mathematics Sch.
Richard Law	Highgate School, London
Alex Lee	Taipei European School
Leslie Leung	Dulwich College
Luozhiyu Lin	Anglo-Chinese School, Singapore
Dmitry Lubyako	Eton College, Windsor
Oleg Malanyuk	English College in Prague
Georgina Majury	Down High School, Co. Down
Stephen Mitchell	St Paul's School, Barnes, London
Yuen Ng	Rainham Mark Grammar School, Gillingham
Alice Rao	Royal High School, Bath
Daniel Remo	Highgate School, London
George Robinson	Brooke Weston Academy, Corby
Edward Rong	Westminster School
Joshua Rowley	Hampton School, Middlesex
Andrew Sellek	Torquay Boys' Grammar School
Zhi Shen	Abingdon School
Alexey Sorokin	Bellerbys College Oxford
Tianyou Tong	Durham School
Kavin Vijayakumar	Bancroft's School, Essex
Sam Watt	Monkton Combe School, Bath
Naomi Wei	City of London Girls' School
Shenyang Wu	Anglo-Chinese School, Singapore
Timothy Xu	Brighton College
Ziming Xue	Anglo-Chinese School, Singapore
Bohan Yu	The Cherwell School, Oxford
Jason Zhang	Bilborough VI Form College, Nottingham
Zigan Zhen	Wellington College, Berkshire
Renzhi Zhou	The Perse School, Cambridge
Zhemin Zhu	Charterhouse, Godalming, Surrey

United Kingdom Mathematics Trust

British Mathematical Olympiad
Round 2: Thursday, 29 January 2015

Time allowed *Three and a half hours.*

Each question is worth 10 marks.

Instructions • *Full written solutions − not just answers − are required, with complete proofs of any assertions you may make. Marks awarded will depend on the clarity of your mathematical presentation. Work in rough first, and then draft your final version carefully before writing up your best attempt.*

*Rough work **should** be handed in, but should be clearly marked.*

• *One or two **complete** solutions will gain far more credit than partial attempts at all four problems.*

• *The use of rulers and compasses is allowed, but calculators and protractors are forbidden.*

• *Staple all the pages neatly together in the top **left**-hand corner, with questions 1, 2, 3, 4 in order, and the cover sheet at the front.*

• To accommodate candidates sitting in other time zones, please do not discuss any aspect of the paper on the internet until 8 am GMT on Friday 30 January.

In early March, twenty students eligible to represent the UK at the International Mathematical Olympiad will be invited to attend the training session to be held at Trinity College, Cambridge (26-30 March 2015). At the training session, students sit a pair of IMO-style papers and eight students will be selected for further training and selection examinations. The UK Team of six for this summer's International Mathematical Olympiad (to be held in Chiang Mai, Thailand, 8-16 July 2015) will then be chosen.

Do not turn over until told to do so.

162

United Kingdom Mathematics Trust

2014/15 British Mathematical Olympiad
Round 2: Thursday, 29 January 2015

1. The first term x_1 of a sequence is 2014. Each subsequent term of the sequence is defined in terms of the previous term. The iterative formula is

$$x_{n+1} = \frac{(\sqrt{2} + 1)x_n - 1}{(\sqrt{2} + 1) + x_n}.$$

Find the 2015 th term x_{2015}.

2. In Oddesdon Primary School there are an odd number of classes. Each class contains an odd number of pupils. One pupil from each class will be chosen to form the school council. Prove that the following two statements are logically equivalent.

    a) There are more ways to form a school council which includes an odd number of boys than ways to form a school council which includes an odd number of girls.

    b) There are an odd number of classes which contain more boys than girls.

3. Two circles touch one another internally at A. A variable chord PQ of the outer circle touches the inner circle. Prove that the locus of the incentre of triangle AQP is another circle touching the given circles at A.

    *The **incentre** of a triangle is the centre of the unique circle which is inside the triangle and touches all three sides. A **locus** is the collection of all points which satisfy a given condition.*

4. Given two points P and Q with integer coordinates, we say that P 'sees' Q if the line segment PQ contains no other points with integer coordinates. An n-loop is a sequence of n points $P_1, P_2, \dots, P_n$, each with integer coordinates, such that the following conditions hold:

    a) P_i sees P_{i+1} for $1 \leqslant i \leqslant n - 1$, and P_n sees P_1;

    b) No P_i sees any P_j apart from those mentioned in (a);

    c) No three of the points lie on the same straight line.

    Does there exist a 100-loop?

The British Mathematical Olympiad 2014-2015
Round 2

The second round of the British Mathematical Olympiad was held on Thursday 29th January 2015. Some of the top scorers from this round were invited to a residential course at Trinity College, Cambridge.

Leading Scorers

40	Yuan Gao	Anglo-Chinese School, Singapore
	Harvey Yau	Ysgol Dyffryn Taf, Carmarthenshire
39	Joe Benton	St Paul's School, Barnes, London
33	Harry Metrebian	Winchester College
31	Samuel Kittle	Simon Langton Boys' Grammar School, Canterbury
30	Andrew Kenyon-Roberts	Aberdeen Grammar School
	Philip Peters	Haberdashers' Aske's School for Boys, Herts
	Askhat Sarkeev	Caterham School, Surrey
29	Xiangjia Kong	Ermysted's Grammar School, N. Yorks
26	Renzhi Zhou	The Perse School, Cambridge
23	Mark Cooper	Horsforth School, Leeds
	Alex Harris	The Perse School, Cambridge
	Jiani Li	Ruthin School, Denbighshire
	Neel Nanda	Latymer School, London
	Marcus Roberts	The Grammar School at Leeds
	Joshua Rowley	Hampton School, Middlesex
22	Joshua Garfinkel	Latymer School, London
	Lawrence Hollom	Churcher's College, Hampshire
	Richard Law	Highgate School, London
	Chris Liu	Winchester College
	Shenyang Wu	Anglo-Chinese School, Singapore
21	Olivia Aaronson	St Paul's Girls' School, Hammersmith
	Kyung Chan Lee	Garden International School, Malaysia
	Thomas Wilkinson	Lambeth Academy

IMO 2015

The 2015 International Mathematical Olympiad took place in Chiang Mai, Thailand, 4-15 July 2015. The Team Leader was Dr Geoff Smith (University of Bath) and the Deputy Leader was Dominic Yeo (Worcester College, Oxford). A full account of the 2015 IMO and the UK preparation for it appears later in the book. The members of the team were Joe Benton, Lawrence Hollom, Sam Kittle, Warren Li, Neel Nanda and Harvey Yau. The reserves were Liam Hughes of Robert Smyth Academy and Harry Metrebian of Winchester College. In addition to the Leader and Deputy Leader, the team were accompanied by Jill Parker, formerly of the University of Bath, to deal with pastoral aspects.

Introduction to the BMO problems and full solutions

The 'official' solutions are the result of many hours work by a large number of people, and have been subjected to many drafts and revisions. The contestants' solutions included here will also have been redrafted several times by the contestants themselves, and also shortened and cleaned up somewhat by the editors. As such, they do not resemble the first jottings, failed ideas and discarded pages of rough work with which any solution is started.

Before looking at the solutions, pupils (and teachers) are encouraged to make a concerted effort to attack the problems themselves. Only by doing so is it possible to develop a feel for the question, to understand where the difficulties lie and why one method of attack is successful while others may fail. Problem solving is a skill that can only be learnt by practice; going straight to the solutions is unlikely to be of any benefit.

It is also important to bear in mind that solutions to Olympiad problems are not marked for elegance. A solution that is completely valid will receive a full score, no matter how long and tortuous it may be. However, elegance has been an important factor influencing our selection of contestants' answers.

Further, from 2014, there was a new annual prize available to entrants of BMO2, in memory of Christopher Bradley. This award, known as the 'Christopher Bradley elegance prize', is to be awarded to the candidate or candidates who, in the opinion of the markers, submitted the most elegant solution or solutions. In 2015 The Christopher Bradley Elegance Prize for an elegant solution to a BMO2 problem was not awarded: there were many pleasant solutions, but none which stood out as exceptional.

BMO Round 1

Problem 1 (Proposed by Julian Gilbey)

Place the following numbers in increasing order of size, and justify your reasoning:

$$3^{3^4}, \; 3^{4^3}, \; 3^{4^4}, \; 4^{3^3} \text{ and } 4^{3^4}.$$

Note that a^{b^c} means $a^{(b^c)}$.

Solution 1 *by Harry Metrebian, Winchester College*:

The five numbers are 3^{81}, 3^{64}, 3^{256}, 4^{27} and 4^{81}. Clearly

$$3^{64} < 3^{81} < 3^{256}.$$

Now $4^{27} = \left(4^3\right)^9 = 64^9$ and $3^{64} = \left(3^4\right)^{16} = 81^{16}$. We can see that $64^9 < 81^{16}$ and so

$$4^{27} < 3^{64} < 3^{81} < 3^{256}.$$

Since $4^{81} > 3^{81}$ it remains to compare 4^{81} and 3^{256}. However $4^{81} = \left(4^3\right)^{27} = 64^{27}$ and $3^{256} = \left(3^4\right)^{64} = 81^{64}$ and it is clear that $64^{27} < 81^{64}$. So

$$4^{27} < 3^{64} < 3^{81} < 4^{81} < 3^{256},$$

that is,

$$4^{3^3} < 3^{4^3} < 3^{3^4} < 4^{3^4} < 3^{4^4}.$$

Solution 2 by Bryan Ng, Bedford School: By calculation, $3^{3^4} = 3^{81}$, $3^{4^3} = 3^{64}$, $3^{4^4} = 3^{256}$, $4^{3^3} = 4^{27}$ and $4^{3^4} = 4^{81}$. We can take the tenth root of all five numbers without affecting the ordering, and since this is equivalent to dividing the exponents by ten, it remains to sort the following numbers:

$$3^{8.1}, \; 3^{6.4}, \; 3^{25.6}, \; 4^{2.7} \text{ and } 4^{8.1}.$$

Then

$$6561 = 3^8 < 3^{8.1} < 3^9 = 19683,$$
$$729 = 3^6 < 3^{6.4} < 3^7 = 2187,$$
$$16 = 4^2 < 4^{2.7} < 4^3 = 64$$
$$65536 = 4^8 < 4^{8.1} < 4^9 = 262144,$$

and

$$262144 < 6561^3 = \left(3^8\right)^3 = 3^{24} < 3^{25.6}.$$

From these we can read off the order:

$$4^{2.7} < 3^{6.4} < 3^{8.1} < 4^{8.1} < 2^{5.6}$$

and so

$$4^{3^3} < 3^{4^3} < 3^{3^4} < 4^{3^4} < 3^{4^4}.$$

Problem 2 (Proposed by Gerry Leversha)

Positive integers p, a and b satisfy the equation $p^2 + a^2 = b^2$. Prove that if p is a prime greater than 3, then a is a multiple of 12 and $2(p + a + 1)$ is a perfect square.

Solution 1 by Harry Metrebian, Winchester College:

Rearranging and factorising, we have

$$p^2 = b^2 - a^2 = (b + a)(b - a).$$

The only positive factors of p^2 are 1, p and p^2. We cannot have that $b + a = b - a = p$ since a is positive, so it must be the case that $b + a = p^2$ and $b - a = 1$. So

$$b = \frac{p^2 + 1}{2} \quad \text{and} \quad a = \frac{p^2 - 1}{2}$$

and therefore

$$2a = p^2 - 1 = (p + 1)(p - 1).$$

Since p is odd, both $p + 1$ and $p - 1$ are even, and moreover one of them is divisible by 4. Hence 8 divides $2a$. Since p is not divisible by 3, one of $p + 1$ and $p - 1$ is divisible by 3 and so 3 divides $2a$. As 8 and 3 are coprime, it follows that 24 divides $2a$ and 12 divides a, as required. Now

$$2(p + a + 1) = 2\left(p + \frac{p^2 - 1}{2} + 1\right)$$

$$= 2p + p^2 - 1 + 2$$

$$= p^2 + 2p + 1$$

$$= (p + 1)^2$$

which is to say that $2(p + a + 1)$ is a perfect square, as required.

Solution 2 by James Roper, Upton Court Grammar School (slightly edited):

Considering the equation modulo 3, and using the fact that b^2 is congruent to either 0 or 1 modulo 3, we see that at least one of p^2 and a^2 is congruent to 0 modulo 3. It cannot be p^2, and therefore $a^j \equiv 0 \bmod 3$ and $p^j \equiv b^2 \equiv 1 \bmod 3$. In particular, 3 divides a.

Considering the equation modulo 4, and using the fact that b^2 is congruent to either 0 or 1 modulo 4, we see that at least one of p^2 and a^2 is congruent to 0 modulo 4. It cannot be p^2, and therefore $a^2 \equiv 0 \mod 4$, and so a is even but b and p are both odd.

Factorising the original equation, $a^2 = (b - p)(b + p)$. Since both b and p are odd, one of $(b - p)$ and $(b + p)$ must be divisible by 4, and since the other is also even, it follows that 8 divides a^2. Therefore 4 divides a. We have shown that a is divisible by both 3 and 4, and since 3 and 4 are coprime, it follows that a is divisible by 12.

Factorising the original equation again, $p^2 = (b - a)(b + a)$. The only positive factors of p^2 are 1, p and p^2. We cannot have that $b + a = b - a = p$ since a is positive, so it must be the case that $b + a = p^2$ and $b - a = 1$. Thus $p + a + 1 = b + p$.

Let us consider all primes q that divide $b + p$. If q is not 2 or p, then q cannot divide $b - p$ (as then it would divide the difference $(b + p) - (b - p) = 2p$) and so q divides $b + p$ exactly as many times as it divides a^2, which is an even number of times. It cannot be the case that p divides $b + p$, as then it would divide both b and a, but $b - a = 1$ and so p would divide 1. We know that exactly one of $(b + p)$ and $(b - p)$ is congruent to 2 modulo 4, and since 2 must divide a^2 an even number of times, 2 divides each of $b + p$ and $b - p$ an odd number of times. So 2 divides $2(b + p)$ an even number of times.

Therefore every prime that divides $2(b + p) = 2(p + a + 1)$ divides it an even number of times. Therefore $2(p + a + 1)$ is a perfect square.

Problem 3 (Proposed by Daniel Griller)

A hotel has ten rooms along each side of a corridor. An olympiad team leader wishes to book seven rooms on the corridor so that no two reserved rooms on the same side of the corridor are adjacent. In how many ways can this be done?

Solution by Agnijo Banerjee, Grove Academy:

Number the rooms on each side of the corridor 1 to 10. We will demonstrate a way in which you can pick n non-adjacent rooms on one side of the corridor.

Pick integers $a_1, a_2, \ldots, a_n$ such that $1 \leqslant a_1 < a_2 < \ldots < a_n \leqslant 11 - n$. Then pick rooms $a_1, a_2 + 1, a_3 + 2, \ldots, a_n + n - 1$. We can see that these room numbers are increasing, and since $a_{i+1} > a_i$, $(a_{i+1} + i) - (a_i + i - 1) \geqslant 2$. So no two such rooms are adjacent.

Conversely, suppose that we have an arrangement of n non-adjacent rooms on one side of the corridor. By ordering them in increasing (numerical) order, and subtracting $(i - 1)$ from the number of the i th room, we get a sequence $b_1, b_2, \ldots, b_n$ with $1 \leqslant b_1 < b_2 < \ldots < b_n = 11 - n$, so all possible arrangements of n non-adjacent rooms arise in the way described above. There are $\binom{11 - n}{n}$ ways to pick $a_1, \ldots, a_n$ and so $\binom{11 - n}{n}$ ways to pick n non-adjacent rooms on one side of the corridor. (Note that it is not possible to pick six or more non-adjacent rooms on one side of the corridor.)

If we label the two sides of the corridor A and B, then we have four choices: picking 5 rooms from A and 2 from B, picking 4 rooms from A and 3 from B, picking 3 rooms from A and 4 rooms from B, or picking 2 rooms from A and 5 from B. Since what we do on one side of the corridor does not affect the other side, there are

$$\binom{6}{5} \cdot \binom{9}{2} + \binom{7}{4} \cdot \binom{8}{3} + \binom{8}{3} \cdot \binom{7}{4} + \binom{9}{2} \cdot \binom{6}{5}$$

$$= 216 + 1960 + 1960 + 216$$

$$= 4352$$

ways of assigning the rooms.

Problem 4 (Proposed by Jeremy King)

Let x be a real number such that $t = x + x^{-1}$ is an integer greater than 2. Prove that $t_n = x^n + x^{-n}$ is an integer for all positive integers n. Determine the values of n for which t divides t_n.

Solution by Harvey Yau, Ysgol Dyffryn Taf:

Firstly, $t_1 = t$ which is an integer. Then $t_2 = x^2 + 2 + x^{-2} - 2 = t^2 - 2$ which is also an integer. More generally,

$$
\begin{aligned}
t_k &= x^k + x^{-k} \\
&= x^k + x^{k-2} + x^{2-k} + x^{-k} - x^{k-2} - x^{2-k} \\
&= \left(x + x^{-1}\right) \cdot \left(x^{k-1} + x^{1-k}\right) - \left(x^{k-2} + x^{2-k}\right) \\
&= t \cdot t_{k-1} - t_{k-2}.
\end{aligned}
$$

Therefore if t_{k-1} and t_{k-2} are both integers, t_k is also an integer. We have shown that t_1 and t_2 are integers, and so it follows by induction that t_n is an integer for all n.

We claim that t divides t_n if and only if n is odd.

Since $t > 2$, we can see that t does not divide t_2. Suppose that k is odd, and t divides t_{k-2} but not t_{k-1}. Then, as $t_k = t \cdot t_{k-1} - t_{k-2}$ it follows that t divides t_k. On the other hand, suppose that k is even, and t divides t_{k-1} but not t_{k-2}. Again, $t_k = t \cdot t_{k-1} - t_{k-2}$, from which it follows that t does not divide t_k. So our claim is true by induction.

Problem 5 (Proposed by David Monk)

Let *ABCD* be a cyclic quadrilateral. Let *F* be the midpoint of the arc *AB* of its circumcircle that does not contain *C* or *D*. Let the lines *DF* and *AC* meet at *P* and the lines *CF* and *BD* meet at *Q*. Prove that the lines *PQ* and *AB* are parallel.

Solution 1 by Sooming Jang, Badminton School:

By the theorem of angles in the same segment, $\angle BCF = \angle BDF$. But since the chord *FA* has the same length as the chord *BF*, the angle subtended by the chord *FA* is equal to the angle subtended by the chord *BF*. So $\angle FCA = \angle BCF$.

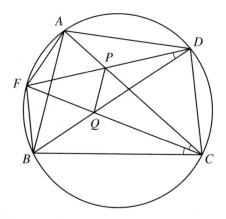

As $\angle QDP = \angle BDF = \angle BCF = \angle FCA = \angle QCP$, it follows by the converse of the theorem of angles in the same segment that quadrilateral *PQCD* is cyclic.

Then $\angle CPQ = \angle CDQ$ by the theorem of angles in the same segment, and $\angle CDQ = \angle CDB = \angle CAB$ by the same theorem. So $\angle CPQ = \angle CAB$. From this it follows, by the converse of the theorem of corresponding angles, that lines *PQ* and *AB* are parallel.

Solution 2 *by Harry Metrebian, Winchester College*: Let *X* be the point of intersection of *AC* and *BD*. Since chords *FA* and *BF* are of equal length, $\angle FCA = \angle FDA = \angle FDB = \angle FCB$ by the theorem of angles in the same segment. We have then that $\angle XDA = \angle XCB$ and $\angle AXD = \angle BXC$ (vertically opposite angles). Therefore triangle *AXD* is similar to triangle *BXC*.

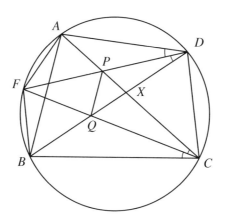

So $\dfrac{XC}{BC} = \dfrac{XD}{AD}$. But also, by the angle bisector theorem in triangles *BXC* and *AXD* respectively, $\dfrac{XQ}{BQ} = \dfrac{XC}{BC}$ and $\dfrac{XP}{PA} = \dfrac{XD}{AD}$. Combining these two, $\dfrac{XQ}{BQ} = \dfrac{XP}{AP}$. Then $\dfrac{BQ}{XQ} = \dfrac{PA}{XP}$ and by adding $1 = \dfrac{XQ}{XQ} = \dfrac{XP}{XP}$ to both sides we have that $\dfrac{XB}{XQ} = \dfrac{XA}{XP}$.

Since $\angle BXA = \angle QXP$, it follows that triangles *BXA* and *QXP* are similar. So $\angle XPQ = \angle XAB$ and so the lines *AB* and *PQ* are parallel by the converse of the theorem of corresponding angles.

Problem 6 (Proposed by Julian Gilbey)

Determine all functions $f(n)$ from the positive integers to the positive integers which satisfy the following condition: whenever a, b and c are positive integers such that $\dfrac{1}{a} + \dfrac{1}{b} = \dfrac{1}{c}$, then

$$\frac{1}{f(a)} + \frac{1}{f(b)} = \frac{1}{f(c)}.$$

Solution by Chikashi Rison, The Leys School:

We claim that $f(mn) = mf(n)$ for all positive integers m and n. We aim to prove this by induction on m. It is clearly true (for all n) when $m = 1$.

Suppose that this is true (for all n) when $m = k$. For some arbitrary n, let $a = (k+1)n$, $b = k(k+1)n$ and $c = kn$. Note that, by factorising, we first have

$$\frac{1}{a} + \frac{1}{b} = \frac{1}{(k+1)n} + \frac{1}{k(k+1)n}$$
$$= \frac{1}{n}\left(\frac{1}{k+1} + \frac{1}{k(k+1)}\right)$$

and then since $\frac{1}{k} = \frac{1}{k+1} + \frac{1}{k(k+1)}$, it follows that

$$\frac{1}{a} + \frac{1}{b} = \frac{1}{n}\cdot\frac{1}{k}$$
$$= \frac{1}{c}$$

and so we can use the condition. Then

$$\frac{1}{f(a)} + \frac{1}{f(b)} = \frac{1}{f(c)}$$

i.e.
$$\frac{1}{f((k+1)n)} + \frac{1}{f(k(k+1)n)} = \frac{1}{f(kn)}$$

i.e.
$$\frac{1}{f((k+1)n)} + \frac{1}{kf((k+1)n)} = \frac{1}{kf(n)}$$

where we use the induction hypothesis to go from the second line to the third. This gives

$$\frac{k+1}{k}\cdot\frac{1}{f((k+1)n)} = \frac{1}{k}\cdot\frac{1}{f(n)}$$

i.e.
$$\frac{1}{f((k+1)n)} = \frac{1}{(k+1)f(n)}$$

and so $f((k+1)n) = (k+1)f(n)$, completing the induction step.

So we have that $f(mn) = mf(n)$ for all positive integers m and n. In particular, $f(m) = mf(1)$ and so any functions f that satisfy this condition are of the form $f(n) = Cn$ for some positive integer constant C. It is clear that all such functions do indeed satisfy the condition.

BMO Round 2

Problem 1 (Proposed by Karthik Tadinada and Dominic Rowland)
The first term x_1 of a sequence is 2014. Each subsequent term of the sequence is defined in terms of the previous term. The iterative formula is

$$x_{n+1} = \frac{(\sqrt{2}+1)x_n - 1}{(\sqrt{2}+1) + x_n}.$$

Find the 2015 th term x_{2015}.

Most successful attempts at this question proceeded to show explicitly that the sequence is periodic. For this case, the terms repeat in blocks of 8. Most students used a careful direct calculation to prove this, while a handful usefully compared the recursive definition to the tangent addition formula. Since the periodicity of the sequence was crucial to obtaining the final answer, it was important to explain this clearly, and not, for example, jump straight from $x_9 = 2014$ to the answer.

Solution 1 *by Olivia Aaronson, St Paul's Girls' School:*
Let $k = \sqrt{2} + 1$ and $m = \sqrt{2} - 1$, so that $mk = 1$ and $k - m = 2$.
This gives

$$x_{n+1} = \frac{kx_n - 1}{x_n + k}$$

$$= \frac{x_n - \frac{1}{k}}{\frac{x_n}{k} + 1}$$

$$= \frac{x_n - m}{mx_n + 1}.$$

Substituting this in, we learn that

$$x_{n+2} = \frac{k\left(\dfrac{x_n - m}{mx_n + 1}\right) - 1}{k + \dfrac{x_n - m}{mx_n + 1}}$$

$$= \frac{x_n - 1}{x_n + 1}$$

after some simplification.

This tells us that $x_{n+4} = \dfrac{-1}{x_n}$ and $x_{n+8} = x_n$. It follows that $x_{n+8k} = x_n$, and so $x_{2015} = x_7$.

Using the results relating x_n to x_{n+4} and x_{n+2}, we see that $x_5 = \dfrac{-1}{2014}$, and so $x_7 = \dfrac{-1 - 2014}{-1 + 2014} = -\dfrac{2015}{2013}$.

Solution 2 by Harvey Yau, Ysgol Dryffyn Taf (slightly edited):
Recall the trigonometric formula

$$\cot(\alpha + \beta) = \frac{\cot(\alpha)\cot(\beta) - 1}{\cot(\beta) + \cot(\alpha)}$$

and that $\cot \frac{\pi}{8} = \sqrt{2} + 1$.

Thus, if we define $\theta_n = \cot^{-1}(x_n)$, we obtain exactly that

$$\cot(\theta_{n+1}) = \frac{\cot\left(\frac{\pi}{8}\right)\cot(\theta_n) - 1}{\cot\left(\frac{\pi}{8}\right) + \cot(\theta_n)} = \cot\left(\theta_n + \frac{\pi}{8}\right).$$

Hence,

$$\theta_{2015} = \theta_1 + \frac{2014\pi}{8} + k\pi$$

$$= \theta_1 - \frac{\pi}{4} + l\pi$$

for some integers k and l, so that

$$x_{2015} = \cot(\theta_{2015}) = \cot\left(\theta_1 - \frac{\pi}{4}\right)$$

$$= \frac{\cot(\theta_1)\cot\left(\frac{\pi}{4}\right) + 1}{\cot\left(\frac{\pi}{4}\right) - \cot(\theta_1)}$$

$$= \frac{2014 + 1}{1 - 2014}$$

$$= -\frac{2015}{2013}.$$

Problem 2 (Proposed by Jeremy King)

In Oddesdon Primary School there are an odd number of classes. Each class contains an odd number of pupils. One pupil from each class will be chosen to form the school council. Prove that the following two statements are logically equivalent.

a) There are more ways to form a school council which includes an odd number of boys than ways to form a school council which includes an odd number of girls.

b) There are an odd number of classes which contain more boys than girls.

There were relatively few successful solutions to this problem. Solutions tended to follow one of two approaches. On the one hand, some proved an explicit formula for the difference between the number of all-boy and all-girl councils, from which the logical equivalence of the statements became clear. On the other hand, some followed a more inductive style of proof, mostly by adding two classes of students at a time. It was in these approaches that students most typically fell down, as in their inductive steps they were only allowed to assume the *equivalence* of the two statements, rather than the *truth* of either.

Solution 1 by Jamie Bell, King Edward VI Five Ways School:

We give boys a value of -1 and girls a value of $+1$. Then, if the product of a council is defined to be the product of its members, a council will have an odd number of boys if and only if its product is negative.

Therefore, the sum of the products of the councils will be the number with an odd number of girls minus the number with an odd number of boys, so we are interested in when this quantity will be positive and when it will be negative. Define the sum of a class to be the sum of its members' values, which is the number of girls minus the number of boys.

Consider the product of the sums of the classes; if we were to write out this product and expand, we would get exactly one term for each council, and thus the product of the sums of the classes is exactly the sum of the products of the councils.

Finally, we observe that there are more ways to form a council with an odd number of boys if and only if this quantity is negative, and this occurs if and only if an odd number of the classes have a negative class sum; that is, an odd number of the classes have more boys than girls.

Solution 2 by Marcus Roberts, The Grammar School at Leeds (slightly edited):

Throughout, let k_E and k_O denote the number of councils with an even and odd number of boys respectively. Let C_B and C_G denote the number of boys and girls respectively in a particular class.

We start by proving that for any school where all classes have more girls than boys there are always more councils with an even number of boys, regardless of whether the number of classes is odd or even. We prove this by induction.

Suppose a school has n classes, each with more girls than boys. If $n = 0$, then there is one council, which is empty and thus has an even number of boys.

For the inductive step, suppose we know that for the first n classes, $k_E > k_O$, and we add a class with $C_G > C_B$. If the new values of the numbers of even and odd councils are k_E' and k_O' respectively, then we have

$$k_E' = C_B k_O + C_G k_E$$

$$k_O' = C_G k_O + C_B k_E$$

and so, by the rearrangement inequality, $k_E' > k_O'$. If, on the other hand, we add a class with more boys than girls, and $k_E > k_O$, the rearrangement inequality tells us that $k_E' < k_O'$. Similarly if we add a class with more girls than boys, and $k_E < k_O$, the same argument shows that $k_E' > k_O'$.

If we construct a school by adding classes one at a time, and there are an odd number of classes with more boys than girls, then we will observe k_E and k_O flipping which is larger an odd number of times, and so at the end $k_E < k_O$. Conversely, if there are an even number of classes with more boys than girls, then the order of k_E and k_O will flip an even number of times and so at the end $k_E > k_O$. This shows that (a) and (b) are logically equivalent.

Problem 3 (Proposed by Gerry Leversha)

Two circles touch one another internally at A. A variable chord PQ of the outer circle touches the inner circle. Prove that the locus of the incentre of triangle AQP is another circle touching the given circles at A.

*The **incentre** of a triangle is the centre of the unique circle which is inside the triangle and touches all three sides. A **locus** is the collection of all points which satisfy a given condition.*

Let I denote the incentre of triangle AQP. In order to determine the locus of I, it seems to be essential to observe that the line AI passes through the contact point (say L) of the chord PQ with the inner circle. The easiest way to discover this fact is to draw an accurate figure. Once this is known empirically, there are several ways to provide a formal justification. After that, there are different ways to proceed, but the various methods have at their heart the ratio $AI : IL$, or something equivalent.

Solution by Joe Benton, St. Paul's School:

Let L be the point where PQ touches ω, and AL meet Ω again at T.

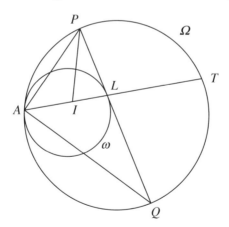

There exists an enlargement taking ω to Ω centred at A which will take L to T. Hence, the tangent to Ω at T is parallel to PQ, so T is the midpoint of arc PQ and thus lies on the angle bisector of $\angle PAQ$. Hence, I lies on AT. It follows that

$$\frac{AI}{IL} = \frac{AP}{PL} \text{ (by the angle bisector theorem)}$$

$$= \frac{TQ}{TL} \text{ (as } APTQ \text{ is cyclic)}$$

$$= \frac{IT}{TL} \text{ (as } T \text{ is the centre of circle } QIP)$$

$$= \frac{AT - AI}{TL}$$

so

$$AI\,(TL + IL) = AT{\cdot}IL.$$

Let $k = \dfrac{AT}{AL}$, which is the scale factor of the enlargement taking ω to Ω. So we have:

$$AI{\cdot}(k - 1)AL + AI{\cdot}IL = kAL{\cdot}IL$$

$$AI{\cdot}(k - 1)AL + AI{\cdot}(AL - AI) = kAL{\cdot}(AL - AI)$$

$$kAI{\cdot}AL - AI^2 = kAL^2 - kAI{\cdot}AL$$

$$kAL^2 - 2kAI{\cdot}AL + AI^2 = 0$$

so, defining $\alpha = \frac{AI}{AL}$, we obtain $\alpha = k \pm \sqrt{k^2 - k}$.

Since $AI < AT$, we obtain that $\alpha < k$, so $\alpha = k - \sqrt{k^2 - k}$ is constant. Hence, I lies on the circle tangent to ω at A which is an enlargement of ω with scale factor α.

Problem 4 (Proposed by Dominic Rowland)

Given two points P and Q with integer coordinates, we say that P 'sees' Q if the line segment PQ contains no other points with integer coordinates. An n-loop is a sequence of n points $P_1, P_2, \ldots, P_n$, each with integer coordinates, such that the following conditions hold:

a) P_i sees P_{i+1} for $1 \leqslant i \leqslant n - 1$, and P_n sees P_1;

b) No P_i sees any P_j apart from those mentioned in (a);

c) No three of the points lie on the same straight line.

Does there exist a 100-loop?

Before attempting this problem, it is useful to observe that two points (a, b) and (c, d) see each other if and only if $a - c$ and $b - d$ are coprime. Only a few candidates submitted complete solutions to this problem, largely following a similar approach to that which is given below.

Solution by Harvey Yau, Ysgol Dyffryn Taf:

On the curve $y = x^2$ no three points are collinear. Since $(a - b)|(a^2 - b^2)$, two integer points (a, a^2) and (b, b^2) will see each other if and only if a and b differ by 1.

Hence, the set of points $\{P_1, P_2, \ldots, P_{99}\}$, $P_j = (j, j^2)$ will satisfy the required properties provided that we can find a suitable P_{100}.

Note that a is coprime to b if and only if a is coprime to $ka + b$, for any integer k.

Since $(-100, 0)$ can see $(1, 1)$ (as their y-coordinates differ by 1), similarly $(-100, 101k)$ must see $(1, 1)$ for any k. Similarly, $(-100, 199k + 99^2 - 1)$ cannot see $(99, 99^2)$.

Now, $(-100, 100^2)$ cannot see any of the points P_2 to P_{98}, so if C is the lowest common multiple of the integers from 102 to 198, $(-100, kC + 100^2)$ will not be able to see any points P_2 to P_{98}.

Since 101 and 199 are prime, they must be coprime to any integer from 102 to 198, and thus also coprime to C. Hence, by the Chinese Remainder theorem there exists an integer N such that:

- $N \equiv 0 \bmod 101$
- $N \equiv 99^2 - 1 \bmod 199$
- $N \equiv 100^2 - 1 \bmod C$.

Then $(-100, N)$ will be a valid choice for P_{100} satisfying conditions (a) and (b).

Finally, we observe that there are an infinite number of choices for N only finitely many of which are eliminated by condition (c), so there is at least one such choice satisfying condition (c).

Olympiad Training and Overseas Competitions

Each year UKMT hold several camps to select and prepare students for participation in the UK team at the International Mathematical Olympiad. Teams are also sent to other international events as part of the training process.

Oxford Training Camp 2014

The Oxford Training Camp was once again held at Queen's College, Oxford. This year 22 students attended the camp which ran from Sunday 24th August until Saturday 30th August. Thanks must go to Peter Neumann who once again organised and directed the camp.

The academic programme was structured and quite intense. There were five 'lecture/tutorial courses' plus some one-off lectures/presentations, all supported by the staff acting as tutors going round and helping the students with the problem-solving activities that were a crucial part of the courses. There were two 100-minute sessions each morning Monday to Friday. These were courses on *Geometry* by Geoff Smith (5 sessions) and *Number theory* by Vicky Neale (5 sessions). On Monday, Tuesday, Thursday and Friday there were three afternoon sessions, each lasting 60 to 75 minutes: sessions on *Problem solving/combinatorics* by Dan Schwarz (4) visiting from Romania, and Paul Russell (4), *Counting Things* by Dominic Yeo (3), and *Inequalities* by Ben Green (1). On the Tuesday evening Ben Green gave an inspirational lecture on *Gaps between primes*. On the final Saturday morning the students sat the four-and-a-half-hour Oxford Mathematical Olympiad under IMO conditions and rules of engagement.

Students attending the Oxford Training Camp this year were: Jaeyon Bae (Headington School), Agnijo Banerjee (Grove Academy), Jacob Coxon (Magdalen College School), Sophie Crane (Stroud High School), Lawrence Hollom (Churcher's College), Xiangjia Kong (Ermysted's Grammar School), Kirsten Land (King's College London Mathematics School), Sichen Liu (Malvern St James), Sophie Maclean (Watford Grammar School for Girls), Georgina Majury (Down High School), Conor Murphy (Eltam College), Michael Ng (Aylesbury Grammar School), Philip Peters (Haberdashers' Aske's School for Boys), Mukul Rathi (Nottingham High School), Alex Song (St Olave's Grammar School), Alan Sun (City of London School), Jacob Sun (Reading School), Alice Vaughan-Williams (Nailsea School), David Veres (King Edward VI School), Naomi Wei (City of London Girls' School), Thomas Wilkinson (Lambeth Academy), Shirley Zhou (Clifton College)

Hungary Camp 2014/15

Once again there was a visit to Hungary over the New Year to train with the Hungarian IMO squad. Twenty British students and 20 Hungarian students attended the camp between 27th December 2014 and 4th January 2015. The group was joined by a student (Luke Gardiner) from the Republic of Ireland again this year.

Thanks go to Dominic Yeo, Roger Patterson and Maria Holdcroft for leading the team this year.

The British students attending were: Hugo Aaronson (St Paul's School), Olivia Aaronson (St Paul's Girls School), Joseph Benton (St Paul's School), Rosie Cates (The Perse School), Sophie Crane (Stroud High School), Alex Harris (The Perse School), Lawrence Hollom (Churcher's College), Liam Hughes (Robert Smyth School), Tomoka Kan (St Paul's Girls School), Samuel Kittle (Simon Langton Boys' Grammar School), Xianghia Kong (Ermysted's Grammar School), Kirsten Land (Kings College London Mathematics School), Warren Li (Eton College), Georgina Majury (Down High School), Bhavik Mehta (Queen Elizabeth's School), Harry Metrebian (Winchester College), Neel Nanda (Latymer School), Michael Ng (Aylesbury Grammar School), Philip Peters (Haberdashers' Aske's School for Boys), Harvey Yau (Ysgol Dyffryn Taf).

Romanian Master of Mathematics

Following the cancellation of The Romanian Master of Mathematics event in 2014, UKMT were pleased to be able to send a team again in 2015. The UK have attended this event since 2008; this year's event took place between 24th February and 2nd March 2015.

This year the team consisted of: Joseph Benton (St Paul's School), Liam Hughes (Robert Smyth School), Samuel Kittle (Simon Langton School), Warren Li (Eton College), Harry Metrebian (Winchester College), Harvey Yao (Ysgol Dyffryn Taf). The team leader was Dominic Yeo and the deputy leader was James Gazet.

The team leader's report for this year can be found at

http://www.imo-register.org.uk/2015-rmm-report.pdf

Balkan Mathematical Olympiad

The 2015 Balkan Mathematical Olympiad was held in Greece between 3rd May and 8th May 2015. The UK is invited to compete in this event as a guest nation. Students are only invited to join this UK team once, so that it gives many more students the opportunity to experience international competition. This year UKMT sent a joint UK/Republic of Ireland team.

The team for 2015 was: Luke Gardiner, Alex Harris (The Perse School), Lawrence Hollom (Churcher's College), Samuel Kittle (Simon Langton School), Kirsten Land (King's College London Mathematics School), Philip Peters (Haberdashers' Aske's School for Boys).

The team was led by Lex Betts and the deputy leader was Gerry Leversha. Jill Parker was in charge of pastoral matters. The leader's report can be read at

http://www.imo-register.org.uk/2015-balkan-report.pdf

Trinity Training Camp

Once again a training and selection camp was held at Trinity College, Cambridge. In 2015 it was held from Thursday 26th March to Monday 30th March. UKMT are very grateful to Trinity College for its continuing support of this camp.

Twenty-two UK students came to Trinity this year. They included all the candidates for this year's IMO team, along with some younger students with potential for future international camps. They were again joined by a student from the Republic of Ireland, Luke Gardiner.

UK attending students were: Olivia Aaronson (St Paul's Girls School), Jamie Bell (King Edward VI Five Ways School), Joseph Benton (St Paul's School), Rosie Cates (The Perse School), Clarissa Costen (Altrincham Girls' Grammar School), Alex Harris (The Perse School), Lawrence Hollom (Churcher's College), Liam Hughes (Robert Smyth School), Andrew Kenyon-Roberts (Aberdeen Grammar School), Samuel Kittle (Simon Langton School), Xianghia Kong (Ermysted's Grammar School), Kirsten Land (King's College London Mathematics School), Warren Li (Eton College), Rory Mclaurin (Hampstead School), Harry Metrebian (Winchester College), Neel Nanda (Latymer School), Philip Peters (Haberdashers' Aske's School for Boys), Thomas Read (The Perse School), Yukuan Tao (Hampton School), Joanna Yass (North London Collegiate School) , Harvey Yau (Ysgol Dyffryn Taf), Renzhi Zhou (The Perse School).

The camp was led by Dominic Yeo (Leader), and Gerry Leversha was deputy leader. Richard Freeland was the local organiser. We are also very grateful to all the other UKMT volunteers who helped at the camp giving sessions and looking after students.

European Girls' Mathematical Olympiad

This year the European Girls' Mathematical Olympiad moved on to Belarus. Since a small beginning in 2012 in Cambridge, when 19 countries participated, the competition in Belarus welcomed 30 countries, including 7 guest countries from outside Europe.

The selection for the UK team was made on performances at BMO1 and BMO2, the two rounds of the British Mathematical Olympiad.

This year's team consisted of: Olivia Aaronson (St Paul's Girls School), Rosie Cates (The Perse School), Kirsten Land (King's College London Mathematics School), Joanna Yass (North London Collegiate School).

The team leader was Jo Harbour and deputy leader was Jenny Owladi.

Tonbridge Selection Camp

The final training camp before selection of the team of six for the IMO was this year held at Tonbridge School from Saturday 23rd May until Wednesday 27th May.

Students participating were: Joseph Benton (St Paul's School), Lawrence Hollom (Churcher's College), Liam Hughes (Robert Smyth School), Samuel Kittle (Simon Langton Boys Grammar School), Warren Li (Eton College), Harry Metrebian (Winchester College), Neel Nanda (Latymer School), Harvey Yau (Ysgol Dyffryn Taf).

Staff at the camp were Geoff Smith, Joseph Myers, Dominic Yeo, and Jill Parker.

The International Mathematical Olympiad

In many ways, a lot of the events and activities described earlier in this book relate to stages that UK IMO team members will go through before they attend an IMO. At this stage, it is worth explaining a little about the structure of the Olympiad, both for its own sake as well as to fit the following report into a wider context.

An IMO is a huge event and takes several years to plan and to execute. In 2015, teams from more than 100 countries went to Chiang Mai, Thailand, to participate. A team consists of six youngsters (although in some cases, a country may send fewer). The focus of an IMO is really the two days on which teams sit the contest papers. The papers are on consecutive days and each lasts $4\frac{1}{2}$ hours. Each paper consists of three problems, and each problem is worth 7 marks. Thus a perfect score for a student is 42/42. The students are ranked according to their personal scores, and the top half receive medals. These are distributed in the ratios gold:silver:bronze = 1:2:3. The host city of the IMO varies from year to year. Detailed contemporary and historical data can be found at

http://www.imo-official.org/

However, whilst these may be the focus, there are other essential stages, in particular the selection of the problems and, in due course, the co-ordination (marking) of scripts and awarding of medals.

As stated, an IMO team is built around the students but they are accompanied by two other very important people: the Team Leader and the Deputy Leader (many teams also take Observers who assist at the various stages and some of these may turn out to be future Leaders). Some three or four days before the actual IMO examinations, the Team Leaders arrive in the host country to deal with the task of constructing the papers. Countries will have submitted questions for consideration over the preceding months and a short list of questions (and, eventually, solutions) are given to Team Leaders on arrival. The Team Leaders gather as a committee (in IMO parlance, the Jury) to select six of the short-listed questions. This can involve some very vigorous debate and pretty tough talking, but it has to be done! Once agreed, the questions are put into the papers and translations produced into as many languages as necessary, sometimes over 50.

At some stage, the students, accompanied by the Deputy Leader, arrive in the host country. As is obvious, there can be no contact with the Team Leader who, by then, has a good idea of the IMO papers! The Leaders and the students are housed in different locations to prevent any contact, casual or otherwise.

On the day before the first examination, there is an Opening Ceremony. This is attended by all those involved (with due regard to security). Immediately after the second day's paper, the marking can begin. It may seem strange that students' scripts are 'marked' by their own Leader and Deputy. In fact, no actual marks or comments of any kind are put on the scripts themselves. Instead, having looked at scripts and decided what marks they think should be awarded, the Leader and Deputy have to justify their claim to others, called co-ordinators, who are supplied by the host country. Once all the marks have been agreed, sometimes after extremely protracted negotiation, the Jury decides where the medal boundaries should go. Naturally, this is a crucial part of the procedure and results in many tears as well as cheers.

Whilst the co-ordination of marks is going on, the students have time to relax and recover. There are often organised activities and excursions and there is much interaction and getting to know like-minded individuals from all corners of the world.

The grand finale is always the closing ceremony which includes the awarding of medals as well as speeches and numerous items of entertainment – some planned but others accidental.

56th International Mathematical Olympiad, Chiang Mai, Thailand, 4-15 July 2015, Report by Geoff Smith (UK Team Leader)

This year the International Mathematical Olympiad was held in Chiang Mai, Thailand. The IMO is the world championship of secondary school mathematics, and is held each July in a host country somewhere in the world. A modern IMO involves more than 100 countries, representing over 90% of the world's population. The competition was founded in 1959. Each participating country may send up to six team members, who must be under 20 years of age and not have entered university.

The UK Deputy Leader was Dominic Yeo of the University of Oxford, and our Observer C was Jill Parker, formerly of the University of Bath. Here is the UK IMO team of 2015.

Joe Benton	St Paul's School, Barnes, London	
Lawrence Hollom	Churcher's College, Petersfield, Hampshire	
Sam Kittle	Simon Langton Boys' Grammar School, Canterbury	
Warren Li	Fulford School, York	
Neel Nanda	Latymer School, Edmonton, London	
Harvey Yau	Ysgol Dyffryn Taf, Carmarthenshire, Wales	

The reserves were Liam Hughes of Robert Smyth Academy and Harry Metrebian of Winchester College.

Here are the results obtained by the UK students this year.

	P1	P2	P3	P4	P5	P6	Σ	Medal
Joe Benton	7	2	1	7	1	1	19	Silver
Lawrence Hollom	7	1	0	1	1	0	10	Honourable Mention
Sam Kittle	7	2	0	7	3	0	19	Silver
Warren Li	7	7	1	7	3	0	25	Silver
Neel Nanda	7	1	0	7	2	0	17	Bronze
Harvey Yau	7	2	1	7	2	0	19	Silver

There are three problems to address on each of two consecutive days. Each exam lasts 4 hours 30 minutes. The cut-offs were 14 for bronze, 19 for silver and 26 for gold. The current IMO marks format became stable in 1981. This is the lowest gold cut, and the equal lowest silver cut, since then. This is evidence of the exceptional difficulty of this IMO, perhaps because of the technical complexity of the medium problems, numbers 2 and 5.

There were 104 teams participating at IMO 2015. Hearty congratulations to the USA for finishing ranked 1st, the first time that they have achieved this since 1994. However, this is the 15th time that they have achieved a top three result in that period, so this is an event which has been waiting to happen. It is very hard to beat a modern Chinese team and the USA joins only the Republic of Korea and Russia in achieving this.

Here are a few of the leading scores (the nations gathering at least 80 points). 1 USA (185), 2 China (181), 3 Korea (161), 4 DPR Korea (156), 5 Vietnam (151), 6 Australia (148), 7 Iran (145), 8 Russia (141), 9 Canada (140), 10 Singapore (139), 11 Ukraine (135), 12 Thailand (134), 13 Romania (132), 14 France (120), 15 Croatia (119), 16 Peru (118), 17 Poland (117), 18 Taiwan (115), 19 Mexico (114), 20 Hungary, Turkey (113), 22 Brazil, Japan, United Kingdom (109), 25 Kazakhstan (105), 26 Armenia (104), 27 Germany (102), 28 Hong Kong (101), 29 Bulgaria,

Indonesia, Italy, Serbia (100), 33 Bangladesh, Slovakia (97), 35 Macao (88), 36 Philippines (87), 37 India (86), 38 Moldova (85), 39 Belarus (84), 40 Israel (83), 41 Saudi Arabia (81), 42 Georgia (80).

Anglophone and Commonwealth interest in other scores might include 49 New Zealand (72), 55 South Africa (68), 57 Malaysia (66), 63 Cyprus (58), 70 Sri Lanka (51), 77 Ireland (37), 82 Trinidad and Tobago (26), 85 Pakistan (25), 88 Nigeria (22).

The inexperienced teams of Botswana, Ghana, Tanzania and Uganda also participated, and as one would expect, did not score heavily. Uganda were only one mark short of getting an honourable mention.

Here are the unusual prizewinners for 2015. The first country to have its rank higher than its score was Algeria. This very creditable performance included a silver medal, and left them only 8 marks behind South Africa. Thus South Africa's position as the traditional champion of Africa may be under threat in the next few years.

It has been a very good year for monarchies, with Australia leading the way in an astonishing 6th place, with Canada in 9th and Thailand in 12th. Australia's result is impressive, especially given their performance when training with the UK at our pre-IMO camp in Malaysia. The teams tied for the Mathematical Ashes (an informal competition between the UK and Australia), and seemed well-matched. However, Australia had a great IMO.

Luxembourg managed to retain the Grand Duchy title, and will keep their firm grip unless Finland or Lithuania revisits its constitutional heritage, or Baden, Mecklenburg-Strelitz or Holstein-Oldenburg breaks away from the Federal Republic of Germany.

Romania was the top member state of the European Union, one spot ahead of France which is the leading country which uses the euro (behind two countries which put Queen Elizabeth on their money: Australia and Canada). France is to be congratulated on finishing ranked above the UK for the first time since 2002.

Syria achieved its first silver medal, and Montenegro its first bronze medal. Trinidad and Tobago obtained only its second ever silver medal. My apologies if I have overlooked other singular achievements.

The Papers

Contestants have 4 hours 30 minutes to sit each paper. The three problems on each paper are each marked out of 7. It is intended that the three problems should be in increasing order of difficulty on each day.

Day 1

Problem 1 We say that a finite set $\mathcal{S}$ of points in the plane is balanced if, for any two different points A and B in $\mathcal{S}$, there is a point C in $\mathcal{S}$ such that $AC = BC$. We say $\mathcal{S}$ that is *centre-free* if for any three different points A, B and C in $\mathcal{S}$, there is no point P in $\mathcal{S}$ such that $PA = PB = PC$.

a. Show that for all integers $n \geqslant 3$, there exists a balanced set consisting of n points.

b. Determine all integers $n \geqslant 3$ for which there exists a balanced centre-free set consisting of n points.

Problem 2 Determine all triples (a, b, c) of positive integers such that each of the numbers

$$ab - c, \qquad bc - a, \qquad ca - b$$

is a power of 2.

(*A power of 2 is an integer of the form 2^n, where n is a non-negative integer.*)

Problem 3 Let ABC be an acute triangle with $AB > AC$. Let Γ be its circumcircle, H its orthocentre, and F the foot of the altitude from A. Let M be the midpoint of BC. Let Q be the point on Γ such that $\angle HQA = 90°$, and let K be the point on Γ such that $\angle HKQ = 90°$. Assume that the points A, B, C, K and Q are all different, and lie on Γ in this order.

Prove that the circumcircles of triangles KQH and FKM are tangent to each other.

Day 2

Problem 4 Triangle ABC has circumcircle Ω and circumcentre O. A circle Γ with centre A intersects the segment BC at points D and E, such that B, D, E and C are all different and lie on line BC in this order. Let F and G be the points of intersection of Γ and Ω, such that A, F, B, C and G lie on Ω in this order. Let K be the second point of intersection of the circumcircle of triangle BDF and the segment AB. Let L be the second point of intersection of the circumcircle of triangle CGE and the segment CA.

Suppose that the lines FK and GL are different and intersect at the point X. Prove that X lies on the line AO.

Problem 5 Let $\mathbb{R}$ be the set of real numbers. Determine all functions $f : \mathbb{R} \to \mathbb{R}$ satisfying the equation

$$f\big(x + f(x + y)\big) + f(y) = x + f(x + y) + yf(x)$$

for all real numbers x and y.

Problem 6 The sequence a_1, a_2, ... of integers satisfies the following conditions:

i. $1 \leqslant a_j \leqslant 2015$ for all $j \geqslant 1$;

ii. $k + a_k \neq \ell + a_\ell$ for all $1 \leqslant k < \ell$.

Prove that there exist two positive integers b and N such that

$$\left| \sum_{j = m + 1}^{n} (a_j - b) \right| \leqslant 1007^2$$

for all integers m and n satisfying $n > m \geqslant N$.

These questions were proposed to the IMO by (1) the Netherlands (Merlijn Staps), (2) Serbia (Dusan Djukic), (3) Ukraine (Danylo Khilko and Mykhailo Plotnikov), (4) Greece (Vaggelis Psychas and Silouanos Brazitikos), (5) Albania (Dorlir Ahmeti) and (6) Australia (Ivan Guo and Ross Atkins).

Forthcoming International Events

This is a summary of the events which are relevant for the UK. Of course there are many other competitions going on in other parts of the world.

The next few IMOs will be held in Hong Kong 2016, Brazil 2017, Romania 2018 and the United Kingdom 2019. Forthcoming editions of the European Girls' Mathematical Olympiad will be in Romania in 2016 and Switzerland in 2017. The Balkan Mathematical Olympiad will be held in Albania in 2016, and the Romanian Master of Mathematics will be in late February 2016.

Diary

This diary is a facetious summary of my personal experience at the IMO and is available on:

www.imo-register.org.uk/2015-report.pdf

Acknowledgements

The UK Mathematics Trust is an astonishing organization, bringing together so many volunteers and a small professional core to focus their energies on maths competitions and more generally, mathematics enrichment. Our collective effort is, I am sure, a significant part of the success story which is secondary school mathematics for able students in the UK. This is not to be complacent, because there are always opportunities do more things and to do things better, but I thank everyone for what we already accomplish every year.

On a personal note, I thank Dominic Yeo, Jill Parker and Joseph Myers for their help during the year and while we were on the road. The teams which UKMT sends abroad to represent the country (and the associated reserves) continue to conduct themselves in an exemplary fashion. We must redouble our efforts to draw in more girls.

I thank Oxford Asset Management for their continuing generous sponsorship of the UK IMO team, and the other donors, both individual and corporate, who give so generously to UKMT. Why not join in?

http://www.ukmt.org.uk/about-us/

G.C.Smith@bath.ac.uk @GeoffBath

UKMT Mentoring Schemes

Following another successful year, the UKMT Mentoring Schemes have again increased in size. These materials are now used in over 1200 schools and with almost 250 individual students working with external mentors. We will continue to offer these free resources to all UK schools and teachers who wish to stimulate and challenge their pupils. We hope that by participating in the schemes, school pupils will be inspired to delve into the subject beyond the curriculum and develop a life-long enthusiasm for mathematics.

The schemes cater for pupils from Years 7 – 13 (in England and Wales, and the equivalent years in Scotland and Northern Ireland). Each participant is linked up with a mentor who can offer help, guidance and encouragement. At the Junior and Intermediate levels we encourage teachers to mentor their own pupils because regular contact is important at this stage, although there are a small number of external mentors available at Intermediate level. At the Senior and Advanced levels, mentees are mentored by undergraduates, postgraduates and teachers who are more familiar with problem-solving techniques, but of course any teacher who is willing to act as a mentor to their own pupils is encouraged to do so. The schemes run from October through to May and anyone who is interested in either being a mentee, a mentor or using the sheets with their classes is welcome to register with Beverley at the UKMT office by emailing mentoring@ukmt.org.uk at any time during the year. Teachers registering on the schemes go onto a mailing list to receive the monthly materials, which they can then use in any way they like with their own students, either individually or in class.

Junior Scheme

Almost 1000 teachers in over 750 schools receive the Junior sheets each month. This scheme caters for those Years 7-9 pupils who have perhaps done well in the Junior Maths Challenge and are looking at Junior Olympiad papers. A few hints are given with the questions which aim to introduce pupils to problem-solving at an accessible level, though the later questions will usually be quite challenging. All pupils are currently mentored by their teachers, and teachers are welcome to enrol in order to use the problem sheets with their classes. This is often a good way to stimulate the interest of a whole class, rather than just one or two individuals, though it is likely that only one or two will rise to producing good solutions to the later questions.

Intermediate scheme

This year the Intermediate scheme was used by around 900 teachers in more than 700 schools. It is aimed at those approximately in Years 9-11 who have done well in the Intermediate Maths Challenge and are preparing for Intermediate Olympiad papers or who have attended one of the UKMT National Mathematics Summer Schools. There is quite a gradient in these problems, from some which can be approached without knowledge of any special techniques to others which require modular arithmetic, some knowledge of number theory and geometrical theorems etc. The aim is to gradually introduce these techniques through the year. As mentees come across these, we hope they will ask questions or look at the internet to find out about these methods. Most pupils are mentored by their teachers, but some external mentors are available where necessary. This year 16 external mentors worked with 47 mentees.

Senior scheme

The senior sheets are sent to more than 450 teachers in 375 schools. Aimed at those in Years 11-13, the questions are set at quite a challenging level, aimed at those who are tackling BMO papers or who have outgrown the Intermediate Scheme. An important role of mentors at this level is to give encouragement to their mentees because the questions are generally more taxing than anything they confront at A-level, and each problem solved is a distinct achievement which should give huge satisfaction.

As well as being used by teachers, the senior scheme also has external mentors available to work with students; typically just one or two students at any school might enrol on the external scheme. This year 76 external mentors helped 173 mentees.

Advanced Scheme

Entry to this scheme is by invitation only as the problems are extremely challenging. It is aimed at UK IMO squad members and others who have outgrown the Senior Scheme, the questions being very hard and mainly of interest to those who are aiming at selection for the UK team in the annual International Mathematical Olympiad (IMO). There were 19 mentees on the scheme this year, working with 9 mentors.

Sample questions from October 2014

The following were questions on the October paper in 2014 – the first paper of the year:

Junior

How many four digit positive integers are there which are multiples of 5 and only use the digits 0, 1, 2, 3, 4 and 5? (Digits may be repeated, and a number may not begin with a zero.)

Intermediate

A postman has seven letters to deliver, one to each of the seven dwarves. Unfortunately, he is rather inaccurate in making his deliveries

(i)   In how many ways could he deliver the letters so that exactly two of the dwarves get the wrong letters?

(ii)   In how many ways could he deliver the letters so that exactly three of the dwarves get the wrong letters?

Senior

A collection of positive integers has a sum of 2014. How large can their product be?

Mentoring conference and dinner

The mentoring conference in November 2014 was held at St Anne's College, Oxford in order to train mentors and exchange ideas and advice. Thanks go to James Cranch, Vicky Neale and Paul Smith for giving sessions.

And finally...

The mentoring schemes are supported by Oxford Asset Management and we would like to thank them for their continuing support.

Our thanks must also go to all the volunteer mentors and question setters who have freely given so much time to make the schemes work and encourage the next generation of young mathematicians. They are too numerous to name here (although they do appear later in the list of volunteers), but without them there would be no schemes at all. If you would like to find out more about becoming a mentor, contact Beverley at UKMT by email: mentoring@ukmt.org.uk

UKMT Team Maths Challenge 2015

Overview

The Team Maths Challenge (TMC) is a national mathematics competition which gives pupils the opportunity to participate in a wide range of mathematical activities and compete against other pupils from schools in their region. The TMC promotes team working and, unlike the Junior, Intermediate and Senior Challenges, students work in groups and are given practical tasks as well as theoretical problems to add another dimension to mathematics.

The TMC is designed for teams of four pupils in:

- Y8 & Y9 (England and Wales)
- S1 & S2 (Scotland)
- Y9 & Y10 (Northern Ireland)

with no more than two pupils from the older year group.

Sample TMC material is available to download from the TMC section of the UKMT website (www.tmc.ukmt.org.uk) for use in school and to help teachers to select a team to represent their school at the Regional Finals.

Report on the 2015 TMC

The thirteenth year of the competition saw yet another record number of participating schools. Entries were received from 1754 teams, of which 1632 turned up to take part at one of 69 Regional Finals.

As usual, competition details and entry forms were sent to schools in early October and made available on the UKMT website, which also provided up-to-date information on Regional Final venues and availability of places, as well as past materials for the use of schools in selecting and preparing their team of four. Schools also received a copy of the winning poster from the 2014 National Final, originally created by Tonbridge Grammar School and professionally reproduced by Arbelos.

Each team signed up to participate in one of the 69 Regional Finals, held between late February and the end of April at a widely-spread set of venues. Each Regional Final comprised four rounds which encouraged the teams to think mathematically in a variety of ways. The Group Round is the only round in which the whole team work together, tackling a set of ten challenging questions. In the Crossnumber the team splits into two pairs; one pair gets the across clues and the other pair gets the down clues. The two pairs then work independently to complete the Crossnumber using logic and deduction. For the Shuttle, teams compete against the clock to answer a series of questions, with each pair working on different questions and the solution of each question dependent on the previous answer. The final round of the day, the Relay, is a fast and lively race involving much movement to answer a series of questions in pairs. Each Regional Final was run by a regional lead coordinator with support from an assistant coordinator and, at some venues, other local helpers. The teachers who accompanied the teams were fully occupied too – they were involved in the delivery and marking of all of the rounds.

TMC National Final

Eighty-eight teams (the winners from each Regional Final plus a few runners-up) were invited to the National Final on 22nd June, which was again held at the Lindley Hall, part of the prestigious Royal Horticultural Halls, in Westminster, London. As usual, the four rounds from the Regional Finals also featured at the National Final except that the Group Round became the Group Circus: a similar round but with the inclusion of practical materials for use in solving the questions. In addition, the day began with the Poster Competition, which is judged and scored separately from the main event. The Poster theme for 2015 was 'Colouring', encouraging a vibrant array of entries which were exhibited down the side of the hall throughout the day for the perusal of the participants as well as the judges.

The following schools, coming from as far north as Fife and as far south as Guernsey, participated at the National Final:

Alleyn's School, London

Ardingly College, West Sussex

Aylesbury Grammar School, Buckinghamshire

Bancroft's School, Essex

Belfast Royal Academy, Belfast

Bethany School, Sheffield

Bishop Bell C of E School, East Sussex

Bolton School (Boys Division), Lancashire

Bristol Grammar School, Bristol

Caistor Grammar School, Lincolnshire

Cardinal Heenan Catholic High School, Leeds

Cargilfield Preparatory, Edinburgh

Cheadle Hulme High School, Cheshire

Cheltenham Ladies' College, Gloucestershire

City of London Freemen's School, Surrey

City of London School

Clifton College, Bristol

Devonport High School for Girls, Devon

Durham Johnston School

Elizabeth College, Guernsey

Ermysted's Grammar School, North Yorkshire

Forest School, London

Glasgow Academy, Glasgow

Gosforth East Middle School, Newcastle upon Tyne

Greenbank High School, Merseyside

Hampton School, Middlesex

Harrow School, Middlesex

Hereford Cathedral School, Hereford

Horsforth School, Leeds

Huntington School, York

Impington Village College, Cambridgeshire

Ipswich School, Suffolk

Kimbolton School, Cambridgeshire

King Edward VI Camp Hill Boys' School, Birmingham

King Edward VI Grammar School, Essex

King Edward VI School, Southampton

King Edward's School, Bath

King Edward's School, Birmingham

Kings' School, Winchester
Lady Manners School, Derbyshire
Lancaster Girls' Grammar School
Lancaster Royal Grammar School
Lancing College, West Sussex
Liverpool Blue Coat School, Merseyside
Loretto School, Midlothian
Loughborough Grammar School, Leicestershire
Magdalen College School, Oxford
Manchester Grammar School
Manchester High School for Girls
Manningtree High School, Essex
Norton Hill School, Bath
Norwich School, Norfolk
Oundle School, Peterborough
Pocklington School, York
Queen Elizabeth's School, Hertfordshire
Reading School, Berkshire
Red House School, Stockton-on-Tees
Royal Grammar School, Guildford, Surrey
Rugby School, Warwickshire
School of St Helen and St Katharine, Oxfordshire
Sevenoaks School, Kent
Sir Roger Manwood's School, Kent
Sir Thomas Picton School, Pembrokeshire

Soar Valley College, Leicester
Solihull School, West Midlands
South Hunsley School, East Yorkshire
South Wilts Grammar School, Wiltshire
St Albans School, Hertfordshire
St Anselm's College, Wirral
St Edward's CoE Academy, Staffordshire
St Faith's School, Cambridgeshire
St Leonard's School, Fife
St Olave's Grammar School, Kent
St Peter's Academy, Staffordshire
Sutton Grammar School for Boys, Surrey
The Fernwood School, Nottingham
The Grange School, Cheshire
The High School of Glasgow, Glasgow
The Nelson Thomlinson School, Cumbria
The Perse School, Cambridge
Tiffin School, Surrey
Torquay Boys' Grammar School, Devon
Truro School, Cornwall
University College School, London
Wellingborough School, Northamptonshire
Westminster Under School, London
Wood Green School, Oxfordshire
Ysgol Friars, Gwynedd

The master of ceremonies was Steve Mulligan, celebrating his tenth year as Chair of the TMC Subtrust and running the event with his usual energy and expertise. In addition, we were delighted to have in attendance Frances Kirwan, Chair of UKMT Council, who addressed the teams and awarded the prizes at the end. We are also grateful to Arbelos for providing additional prizes for the event and to UKMT volunteer Andrew Bell for capturing the day's excitement in his additional role as official photographer. Congratulations go to the 2015 Team Maths Challenge champions Westminster Under School (London) and to the winners of the Poster Competition: The Perse School (Cambridge).

As usual, thanks are due to a great number of people for ensuring another successful year of the TMC: the team of volunteers (listed at the back of this book) who generously give up their time to write, check and refine materials, run Regional Finals (with a helping hand from family members in a few cases!) and readily carry out countless other jobs behind the scenes; the staff in the UKMT office in Leeds for the way in which the competition is administered (particularly Nicky Bray who has responsibility for the central coordination of the competition, assisted by Shona Raffle-Edwards and Jo Williams, with additional support from

Gerard Cummings) and the team of packers for their efficient and precise preparation and packing of materials; the teachers who continue to support the competition and take part so willingly, some of whom also undertake the significant task of organising and hosting a Regional Final at their own school and, of course, the pupils who participate so enthusiastically in the competition at all levels. Our thanks also go to additional contacts at schools and other host venues responsible for organising and helping with Regional Finals (listed at the back of this book).

TMC Regional Finals Material

Each of the 69 Regional Finals held across the UK involved four rounds:
1. Group Round   2. Crossnumber
3. Shuttle   4. Relay Race

Group Round
Teams are given a set of 10 questions, which they should divide up among themselves so that they can answer, individually or in pairs, as many as possible in the allotted time.

Question 1
The mean age of four pupils, on the date of a TMC Regional Final they had been entered for, was 14 years 2 months.

Their teacher realised that this date clashed with school holidays and changed their entry to take place exactly two months later.

Unfortunately, this meant that one of the pupils could no longer participate and a younger replacement was found.

The mean age of the revised team of four pupils, on the new entry date, was 13 years 11 months.

How many months older is the original entrant than the younger replacement?

Question 2
(a) What is the smallest sum of three *different* prime numbers, each of which is the sum of two *different* prime numbers?
(b) What is the smallest sum of three *different* prime numbers, each of which is the sum of three *different* prime numbers?

Question 3

The vowels A, E, I, O and U are coded by the squares 1, 4, 9, 16 and 25 respectively.

The consonants B, C, D, . . . , X, Y and Z are coded by the integers less than or equal to 26 that are not squares, in reverse. So B = 26, C = 24, D = 23, ..., X = 5, Y = 3 and Z = 2.

(a) Decode the message

$$20\ 16\ 6 \qquad 1\ 11\ 4 \qquad 3\ 16\ 25$$

(b) Decode the message

$$7\ 4\ 11\ 3 \qquad 6\ 4\ 17\ 17 \qquad 8\ 20\ 1\ 14\ 18\ 10$$

Question 4

In an isosceles trapezium *ABCD*, the lines *BC* and *AD* are parallel and *AB* = *CD*. The point *E* lies on *DA* such that

$$AB\ =\ BE$$

$$\text{and} \qquad \angle BCD\ =\ 3 \times \angle CDE.$$

What is the value of $\angle ABE$ in degrees?

Question 5

In the following equation, *a*, *b* and *c* are different digits, and '*bc*' and '*cb*' are two-digit numbers, such that

$$a \times \text{'}bc\text{'} \times \text{'}cb\text{'}\ =\ 2015.$$

What is the value of $a + b + c$?

Question 6

Andrew drove one and a half miles to work and then walked back home at a tenth of his average driving speed.

The two journeys lasted a total of 33 minutes.

What was Andrew's average walking speed in mph?

Question 7

At a puzzle conference 68% of those attending enjoyed doing crosswords and 39% enjoyed doing Sudoku.

Only 3% of those attending the conference did not enjoy either doing crosswords or doing Sudoku.

What percentage of those attending the conference enjoyed both of these activities?

Question 8

The order of the digits is reversed in a certain two-digit whole number.

This gives a new whole number which is one less than half of the original number.

What is the original number?

Question 9

A rectangle *ABCD* is cut into two regions by a single straight line from the corner *B* to a point *E* on the opposite side *AD*.

Let the length of *AE* be *x* cm and let the length of *DE* be *y* cm.

(a) The areas of the regions *ABE* and *BCDE* are in the ratio 1 : 2. What is the ratio *x* : *y*?

(b) Suppose the ratio *x* : *y* is changed to 6 : 5.

What is the ratio

area of triangle *ABE* : area of quadrilateral *BCDE*?

Question 10

A woman says to her brother "I am four times as old as you were when I was the same age as you are now."

The woman is 40 years old.

How old is her brother now?

Crossnumber

Teams are divided into pairs, with one pair given the across clues and one pair given the down clues. Each pair answers as many questions as possible on the grid, showing their answers to the supervising teacher who either confirms or corrects them. The correct version is then shown to both pairs. The sole communication permitted between the two pairs is to request, via the supervising teacher, for a particular clue to be solved by the other pair.

Across:

2.	The smallest 4-digit palindromic cube	(4)
6.	A square plus its square root	(3)
8.	A number equal to the sum of the cubes of its digits	(3)
9.	$\sqrt{13^2 - 5^2}$	(2)
10.	A factor of 10 Down	(2)
11.	The remainder when 26 Across is divided by 22 Across	(3)
13.	Four more than 1 Down	(2)
14.	A multiple of 17 Across	(2)

16. The square of a prime factor of 2015 (2)
17. The highest common factor of 26 Across and 2015 (2)
18. 14 Across plus 25 Down (3)
20. A Fibonacci number (2)
21. 23 Down plus a non-trivial factor of 18 Down (2)
22. A multiple of 23 Down (3)
24. x where $4 = 5(x - 2) - 26$ Across (3)
26. 6 Across plus 15 Down (4)

Down:

1. The largest prime factor of 2015 (2)
3. A multiple of 9 Across (2)
4. A Fibonacci number (3)
5. $9^3 + 10^3$ or $12^3 + 1^3$ (4)
6. The sum of its digits multiplied by an even number (3)
7. A cube (3)
10. A multiple of 10 Across (3)
12. The product of 13 Across and 16 Across (3)
14. A number equal to the sum of the cubes of its digits (3)
15. The mean of 6 Across and 7 Down (3)
16. 2 Across plus 26 Across (4)
18. A multiple of 11 (3)
19. 14 Down minus 9 Across (3)
20. A number that makes the sum of the 6 digits in its column
equal to 17 Across (3)
23. The product of 17 Across and a square (2)
25. Both a square and a cube (2)

Shuttle

Teams are divided into pairs, with one pair given Questions 1 and 3 (along with the record sheet on which to record their answers) and the other pair given Questions 2 and 4. The first pair works on Question 1 and then passes the answer to the students in the other pair who use it to help them answer Question 2, for which they can first carry out some preparatory work. This continues with the second pair passing the answer to Question 2 back to the first pair and so on until a full set of answers is presented for marking. Bonus points are awarded to all teams which present a correct set of answers before the 6-minute whistle, then the other teams have a further 2 minutes in which to finish. Four of these shuttles are attempted in the time given.

A1 $A = (12 \times 11 \times 10) - (9 + 8 + 7) + (6 \times 5 \times 4 \times 3 \times 2) - 1$

Pass on the sum of the digits of A.

A2 *T is the number that you will receive.*

A cuboid has a square base measuring 12 cm by 12 cm. The cuboid has volume $108T$ cm^3 and total surface area A cm^2.

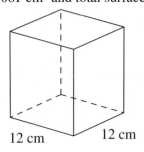

12 cm   12 cm

Pass on the value of A.

A3 *T is the number that you will receive.*

In the shape shown,
$CD = EC = EB = EA$.

Angle BEA is $\frac{1}{6}T°$ and angle CDE is $K°$.

Pass on the value of K.

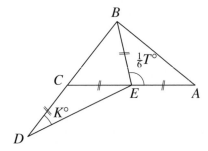

A4   *T is the number that you will receive.*

Dr. Prime, the headmaster of Maths Academy, likes to have all his classes the same size.

Last year, when there were fewer than 600 pupils in his school, he noticed that he could divide them exactly into classes of 36 pupils, 30 pupils or *T* pupils.

Write down the number of pupils in Maths Academy.

B1
$$B = \frac{0.9 \times 0.08 - 0.7 \times 0.06}{0.005}$$

Pass on the value of *B*.

B2   *T is the number that you will receive.*

A is the total number of edges of a prism whose cross-section is a *T*-sided polygon.

B is the total number of faces of a pyramid whose base is a $(T - 1)$-sided polygon.

Pass on the value of $A - B$.

B3   *T is the number that you will receive.*

T and $(2T - 10)$ are the first two terms in a linear sequence.

The 20 th term of the sequence is *K*.

Pass on the value of *K*.

B4   *T is the number that you will receive.*

The diagram shows lines with equations $y = x$, $y = T$, $x = 2T$ and $x = 3T$.

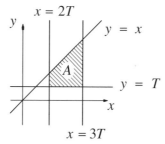

Write down the value of the area of region *A*.

C1
$$C = \left(6\tfrac{6}{7} \times 7\tfrac{7}{8} \times 8\tfrac{8}{9}\right) \div \left(2\tfrac{2}{3} \times 3\tfrac{3}{4}\right)$$

Pass on the value of C.

C2   *T is the number that you will receive.*

Adam is now twice as old as Zoe.

In two years' time, Adam will be four times as old as Zoe was one year ago.

In T years' time, Adam will be X years old.

Pass on the value of X.

C3   *T is the number that you will receive.*

In Mathstown Academy, the T people in year 7 were asked if they like Rugby and also if they like Cricket.

$\tfrac{1}{2}T$ said they like Rugby.

$\tfrac{2}{3}T$ said they like Cricket.

$\tfrac{1}{9}T$ said they like neither.

Pass on how many like both sports.

C4   *T is the number that you will receive.*

I am thinking of a positive whole number. I perform the following calculations on it (in order).

multiply by 4

add 7

multiply by 3

subtract 27

divide by 6

add 4

The answer is T.

Write down my original number.

D1
$$D = \frac{9^2 - 8^2 + 7^2 - 6^2 + 5^2 - 4^2 + 3^2 - 2^2 + 1^2}{9 - 8 + 7 - 6 + 5 - 4 + 3 - 2 + 1}$$

Pass on the value of D.

D2 *T is the number that you will receive.*

$$Tx - 2y = 8$$

$$x + 6y = 3T + 5$$

Solve the simultaneous equations and pass on the value of x.

D3 *T is the number that you will receive.*

Flora likes to invest her money in shares. At the start of 2012, she invested £1000.

The value of her investment increased by $10T\%$ in 2012, increased by £$100T$ during 2013, but decreased by $10T\%$ during 2014.

By the end of 2014, the value of Flora's total investment differed by £F from the original value.

Pass on the value of F.

D4 *T is the number that you will receive.*

The human brain is estimated to be able to store 2.4 petabytes of information.

A DVD holds about 5 gigabytes of information.

It takes 1 minute to burn a DVD of information. You have T DVD-burners all working at the same time. Write down how long it would take to burn your entire brain of data to DVDs. Give your answer to the nearest hour.

[1 *petabyte* $\approx 10^{15}$ *bytes, and* 1 *gigabyte* $\approx 10^9$ *bytes.*]

Relay

The aim here is to have a speed competition with students working in pairs to answer alternate questions. Each team is divided into two pairs, with each pair seated at a different desk away from the other pair and their supervising teacher.

One member of Pair A from a team collects question A1 from the supervising teacher and returns to his/her partner to answer the question together. When the pair is certain that they have answered the question, the runner returns to the front and submits their answer. If it is correct, the runner is given question B1 to give to the other pair (Pair B) from their team. If it is incorrect, Pair A then has a second (and final) attempt at answering the question, then the runner returns to the front to receive question B1 to deliver to pair B. The runner then returns, empty handed, to his/her partner. Pair B answers question B1 and a runner from this pair brings the answer to the front, as above, then takes question A2 to Pair A. Pair A answers question A2, their runner returns it to the front and collects question B2 for the other pair, and so on until all questions are answered or time runs out. Thus the A pairs answer only A questions and the B pairs answer only B questions. Only one pair from a team should be working on a question at any time and each pair must work independently of the other.

A1   Harry eats 35% of a cake, Martha eats 15% of it and Joshua, who does not understand percentages yet, eats one fifth of it.

What fraction is left for their grandmothers to eat?

A2   A milk tanker travels 40 miles from the farm to the cheese factory at a constant speed of 60 mph, starting with a full tank of 400 gallons of milk. The milk has been leaking at a constant rate throughout the journey at 75 gallons per hour.

How much milk remains in the tank when it arrives?

A3   A litre of water weighs 1 kg, and a litre of ice weighs 870 g.

What is the difference in kg between the weights of 6 litres of water and 5 litres of ice?

A4   What is the smallest positive integer x for which $x^2 + 3x + 5$ is a prime number?

A5   The integers from one to eleven are written in numerical order. They are then rewritten in alphabetical order.

Two integers each appear in the same position in the alphabetical list as they do in the numerical list.

What is their sum?

A6 I was interested in studying the word length of a passage in the book I am reading. Below is a frequency table of my findings.

Number of letters	1	2	3	4	5	6	7	8	9
Frequency	0	1	2	3	2	0	1	0	1

What is the number you get by multiplying the mean by the range?

A7 A rectangle has an area of 66 cm².

When the height of the rectangle is increased by 2 cm and the length decreased by 3 cm a square is formed.

What is the area of the square?

A8 In the diagram, O is the centre of the circle, and A, B and C are points on the circumference.

Angle $OAB = 70°$ and angle $OCB = 40°$.

What is the angle ABC?

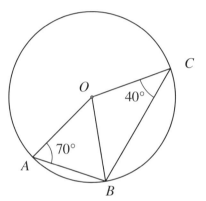

A9 A shop sells zigs, zogs and zags.

Five zigs cost the same as six zags, and seven zogs cost the same as four zigs.

Write them in order of cost, cheapest first.

A10 What is the median of all the factors of 2015?

A11 A metal bar weighs 68 kg, and it is made into 20 000 identical pendants.

What is the weight of each pendant in grams?

A12

115	116	117	118
135	136	137	138
155	156	157	158
175	176	177	178

What is the sum of all the prime numbers in the table above?

208

A13 What is the difference in the acute angles between the hands of a clock showing two o'clock and between the hands of a clock showing quarter past two?

A14 I have a square garden of side 6.4 m to make into a vegetable patch. I decided to put sleepers onto the square to form a frame around the garden. Each sleeper is 60 cm wide.
What will be the surface area of the soil inside the sleeper frame?

$\longmapsto$ 6.4 m $\longrightarrow$

A15 Three of the vertices of a rectangle are $(-3, -3)$, $(5, 1)$ and $(-1, -5)$.
What are the coordinates of the centre of the rectangle?

B1 40% of the members of a school mathematics club were boys, and there were 72 girls.
How many members did the club have?

B2 a and b are positive integers such that $a^2 + b^2 = 13^2$.
What is the value of ab?

B3 I ran a race at a constant speed and my time was recorded as 4.80 minutes.
Change this time to seconds.

B4 What is the difference in the acute angles between the hands of a clock showing one o'clock and between the hands of a clock showing twenty past one?

B5 A block of plastic weighing 60 kg is made into 15 000 identical key fobs.
What is the weight of each fob in grams?

B6 A shop sells migs, mogs and mags.
Five migs cost the same as six mags, and seven mogs cost the same as four migs.
Write them in order with the cheapest first.

B7 What is the smallest positive integer x for which $x^2 + 2x + 7$ is a prime number?

B8   What is the sum of the median, mode, mean and range of the following set of numbers?

$$3, \quad 3, \quad 4, \quad 5, \quad 10.$$

B9   A quadrilateral can have 1, 2 or 4 interior right angles.

What is the greatest possible number of interior right angles in a hexagon with all angles less than 180°?

B10   The area of a rectangle is 238 cm².

When the height of the rectangle is increased by 1 cm and the length decreased by 2 cm a square is formed.

What is the area of the square?

B11   The rectangle *ABCD* is reflected in the line $y = x$.

$A = (10, 2), B = (8, 2), C = (8, 7), D = (10, 7).$

What are the coordinates of the centre of the new rectangle after reflection?

B12   I tried to draw a kite without measuring instruments.

I measure the angles on my shape and notice that it is not quite symmetrical.

Angle *BAC* = 35°, angle *ABC* = 120°, angle *ADC* = 130° and angle *CAD* = 30°.

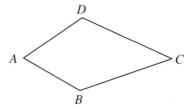

What is the reflex angle at *C*?

B13   Art ate $\frac{1}{3}$ of a packet of sweets, Bert ate 25% and Caz ate $\frac{1}{5}$.
What fraction was left for Diz?

B14   What is the mean of all the factors of 2015?

B15   Speedy Sam drives at an average speed of 66 mph on a trip, and cautious Chris drives back at an average speed of 44 mph.
What is their average speed for the whole trip?

TMC National Final Material

At the National Final, the Group Round is replaced by the Group Circus.

Group Circus

Teams move around a number of stations (eight at the 2015 National Final) to tackle a variety of activities, some of which involve practical materials.

Station 1

$P = (n + 1)^3 - n^3$, where n takes the values 0, 1, 2, 3, 4, 5, 6, 7, 8 in turn.

Find the total of all the values of P which are prime numbers.

Station 2

(a) On the grid below draw all the different straight lines that pass through *exactly two* points.

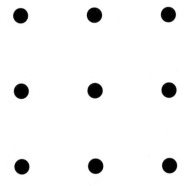

(b) You are provided with nine counters arranged in a 3 × 3 square grid (as shown). Move one counter only so that nine different straight lines can be drawn on the grid each passing through *exactly three* counters.

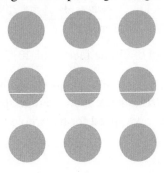

Station 3

In the equations below the letters A, B, C and D represent different non-negative single-digit numbers.

$$D = \tfrac{1}{2}(A - B)$$
$$D^2 = A - B$$
$$D^3 = A + B + C + D$$

'$ABCD$' stands for the four-digit number formed by replacing each of A, B, C and D by a single digit.

What is the four-digit number '$ABCD$'?

Station 4

(a) You are provided with nine lollipop sticks. Arrange the nine sticks to form five equilateral triangles. You are not allowed to place sticks over others.

(b) Arrange eight of the sticks to form a figure which contains two squares, eight triangles and an eight-pointed star. You are allowed to place sticks over others.

Station 5

Three regular polygons, with different numbers of sides, fit together at a point. Their angles all contain a whole number of degrees.

Two examples of three such regular polygons, showing the numbers of sides, are shown.

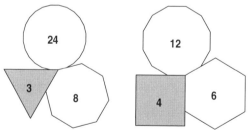

There are three other possible cases. The smallest interior angle in each case is either 60° or 90°.

What are the two largest interior angles in these three cases?

NOTE: The interior angle of an n-sided regular polygon is $180° - \dfrac{360°}{n}$.

212

Station 6

You are provided with a pentagon made out of card. The pentagon consists of a square with a right-angled isosceles triangle placed on top of it (as shown).

You are required to cut the pentagon into three pieces only and to then rearrange the pieces to produce a right-angled isosceles triangle. Your two cuts must be along completely straight lines and the first one must be along the dashed line shown.

The three pieces of card should not overlap, and there should be no gaps between them.

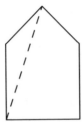

Station 7

Using the five digits 1, 2, 3, 4 and 5 once each it is possible to make 120 different five-digit whole numbers ranging from 12345 to 54321.

All 120 of these different five-digit numbers are added up.

What is the total?

Station 8

You are provided with the number cards 1-9. Place one of the cards over each square in the large triangle below.

The four numbers along each side of the triangle must add up to the same total.

There is more than one possible solution, but you only need to find one.

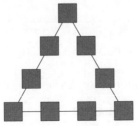

Crossnumber

Across:

1.	A factor of 3 Across	(3)
3.	13 Down plus 16 Across	(3)
5.	16 Across multiplied by 10^2 plus 13 Down	(5)
7.	A cube number	(3)
9.	A Fibonacci number	(3)
10.	The product of five consecutive prime numbers	(4)
12.	Twice the sum of the number of vertices and edges of a cube	(2)
13.	The highest common factor of 10 Across and 14 Down	(2)
14.	The mean of 3 Down and 10 Across	(4)
16.	A factor of 10 Across	(3)
17.	The sum of ten consecutive Fibonacci numbers	(3)
19.	The product of the digits of this palindromic number is 2^{10}	(5)

20. The difference between 10 Across and 14 Down   (3)

21. This number is the sum of all the 2-digit numbers made
    from its digits   (3)

Down:

1. The product of the digits of 9 Across   (3)

2. The sum of the first ten prime numbers   (3)

3. The sum of 10 Across and 14 Down   (4)

4. The sum of its digits is equal to 13 Across   (5)

6. 20 Across minus 9 Across   (3)

8. Seven multiplied by 12 Across   (3)

10. Twice a square Fibonacci number   (3)

11. The difference between 16 Across and 17 Down   (3)

12. The product of the digits of this number is 2^6 (5)

13. A square number   (3)

14. The product of seven consecutive Fibonacci numbers   (4)

15. Twice 9 Across   (3)

17. A palindromic square   (3)

18. The product of the sum of the digits in the third column
    and the sum of the digits in the seventh row   (3)

Shuttle

A1   In the diagram, lines *AD* and *EH* are parallel. The lengths of *FB* and *FG* are equal and the lengths of *BG* and *GC* are equal. Angle *ABF* is 50°.

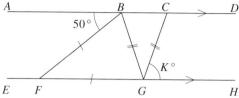

Angle *CGH* = *K*°.

Pass on the sum of the digits of *K*.

A2   *T is the number that you will receive.*

The diagram shows a circle of radius $(T + 1)$ cm and a smaller circle of radius 4 cm, with the same centre.

The area between the circles is divided into six equal regions.

The shaded area is $k^2\pi$ cm².

Pass on the value of *k*.

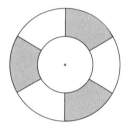

A3   *T is the number that you will receive.*

Solve the equation

$$2T(x - 5) - 3(T + 1 - x) = 5(2x - 3) - 2(x + 3T).$$

Pass on the value of *x*.

A4   *T is the number that you will receive.*

The expression $12^{T+1} \times 30^{T-1}$ can be written as $2^a \times 3^b \times 5^c$, where *a*, *b* and *c* are integers.

Write down the value of $a + b + c$.

B1 U is a prime number, K is a cube, M is a square and T is a triangular number.

U, K, M and T are all different and smaller than 10.

Also, $U \times K = M \times T$.

Pass on the value of $U + K + M + T$.

B2 *T is the number that you will receive.*

The first two terms of a sequence are 3 and 1, respectively, and then each successive term is the sum of the previous two.

The Tth term ends in the digit k.

Pass on the value of k.

B3 *T is the number that you will receive.*

The mean of these numbers (which are in ascending order)

$$4, \quad 6, \quad T, \quad T + 2, \quad 14, \quad x$$

is three times the median.

Pass on the value of $\dfrac{x}{10}$.

B4 *T is the number that you will receive.*

The ratio of the area of shape A to the area of shape B is 3 : 4.

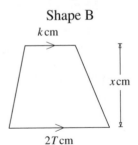

Shape A

x cm

$2T$ cm

Shape B

k cm

x cm

$2T$ cm

Write down the value of k.

C1   Pass on the value of

$$\left(1 - \frac{1}{2}\right)\left(2 - \frac{2}{3}\right)\left(3 - \frac{3}{4}\right)\left(4 - \frac{4}{5}\right)\left(5 - \frac{5}{6}\right).$$

C2   *T is the number that you will receive.*

Candy started the week with T sweets.
    On Monday she ate a quarter of them.
    On Tuesday she chomped a fifth of the remainder.
    On Wednesday she gulped down a third of what was left.
This left Candy with S sweets for the rest of the week.
Pass on the value of S.

C3   *T is the number that you will receive.*

Let $U = \dfrac{1}{K - 1}$, $K = \dfrac{1}{1 - M}$ and $M = \dfrac{1}{T - 1}$.
Pass on the value of U.

C4   *T is the number that you will receive.*

Anthony the ant leaves his home in the Sahara Desert and walks 2.5 metres north, then 4 metres east, before stumbling T metres south, and finally 16 metres west.
Write down how many metres Anthony is now from home.

D1   Melody had a cube of side length 4 cm. From the centre of each face she cut out a cube of side length 1 cm, resulting in a solid shape with volume 58 cm^3.
The total surface area of her solid shape is now K cm^2.
Pass on the value of $\dfrac{K}{10}$.

D2   *T is the number that you will receive.*

In a certain class, 75% of those who like tofu are vegetarian, but only half of those who are vegetarian like tofu.
T% of the class like tofu, and K% of the class is vegetarian.
Pass on the value of K.

218

D3   *T is the number that you will receive.*

The diagram below shows a rectangle *ABCD*. The ratio *EF* : *FB* is 1 : 3.

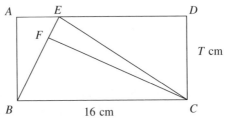

The area of triangle *CEF* is *K* cm².
Pass on the value of *K*.

D4   *T is the number that you will receive.*

Karloff is exactly *T* years old.
When a third of Karloff's age is added to half of Boris's age, the result is the difference between Boris's and Karloff's ages.
Write down the youngest that Boris could be.

Relay

A1   What is the median of these four numbers?

20 + 15	20 − 15
20 × 15	20 ÷ 15

A2   Each term of a sequence is formed by doubling the previous term and adding 2.
The first term is 3.
What is the smallest three-digit number in the sequence?

A3   A mathematical grandmother is making jam, but finds she only has 600 g of sugar, instead of the one kilogram that the recipe needs for 1.25 kg of plums.
What weight of plums, in kilograms, must she use for the reduced amount of jam?

A4   A tennis coach packs tennis balls into square boxes holding 9 balls, and hexagonal ones holding 7 balls.
What is the smallest number of boxes she can use to contain exactly 100 balls, with no spaces?

A5 A circle has radii *OA* and *OB* which divide it into two sectors whose areas are in the ratio 4 : 11.

What is the size of angle *OAB* in triangle *AOB*?

A6 What are the co-ordinates of the point where the lines $y = x - 2$ and $y = 2x - 4$ cross?

A7 Shona glances at her neighbour's digital watch at the start of the chairman's speech and writes down the time as 12:21.

At the end of the speech she is surprised to see the watch showing 95:21, and realises she has been looking at it upside down.

How long in minutes was the chairman's speech?

A8

In the diagram, $a = 132$.
What is the value of *b*?

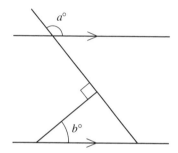

A9 What is the ratio of the number of primes between 30 and 60 to the number of squares between 30 and 60?

A10 A health centre recorded 380 patients visiting the Accident and Emergency Department at a total cost of £36 091.

To the nearest pound, what is the average cost per patient?

A11 The total of 13 consecutive integers is 2015.
What is the largest of the integers?

A12 Write these improper fractions in ascending order:

$$\frac{95}{12}, \quad \frac{79}{10}, \quad \frac{87}{11}, \quad \frac{71}{9}.$$

A13 Posters measuring 50 cm by 75 cm must be displayed in portrait orientation. The maximum possible number of posters are put up on a board measuring 1.8 m wide and 0.9 m high.

The posters must not overlap.

What is the area, in cm², of board left uncovered?

A14   In the diagram, the areas of the shaded rectangles are equal. What is the value of x?

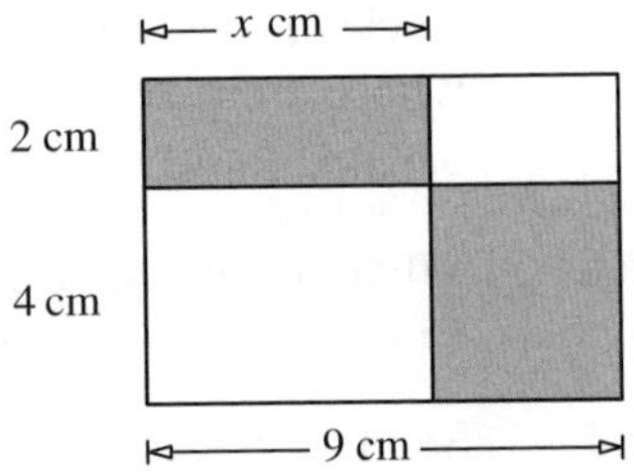

A15   Friends Arthur, Bert, Claire, Dave, Emma, and Fred buy the last 6 tickets for the play 'Much Ado About Zero'.

Unfortunately they are not all together: there are three together on row G; two next to each other on row H; and one on their own on row J. The order within these rows is not significant.

Emma and Fred refuse to sit on their own.

How many ways can the groups be organised to sit in the rows they have been allocated?

B1   A tennis coach packs tennis balls into rectangular boxes, each holding 6 balls, and hexagonal ones, each holding 7 balls.

What is the smallest number of boxes she can use to contain exactly 100 balls, with no spaces?

B2   A pile of 250 cards is 1 m high.

How thick in millimetres is each card?

B3   What are the co-ordinates of the point where the lines $y = x + 4$ and $y = 2x + 1$ cross?

B4   At the Polynumeral Theme Park adult tickets are £8.90 and a child's ticket (aged 5 – 13) is £6.50. Children under 5 years are free. A family ticket (maximum of 2 adults and 3 children 5-13) is only £30.

What will be the lowest price for 3 adults and 4 12-year-old children?

B5   Each term of a sequence is formed by doubling the previous term and subtracting 1.

The first term is 4.

What is the smallest three-digit number in the sequence?

B6   Write these improper fractions in ascending order:

$$\frac{100}{9}, \quad \frac{78}{7}, \quad \frac{89}{8}, \quad \frac{111}{10}.$$

B7   A health centre recorded 281 appointments in one month where patients failed to arrive.
The average length of an appointment is 5 minutes.
To the nearest hour, how much time was wasted?

B8   A circle has radii *OA* and *OB* which divide it into two sectors whose areas are in the ratio 7 : 11.
What is the size of angle *OAB* in triangle *AOB*?

B9   The total of 31 consecutive integers is 2015.
What is the largest?

B10   Four grandchildren in a room are 2, 3, 5 and 10 years old.
The mean age of the grandchildren in the room increases by 2 years when a fifth grandchild enters.
How old is the fifth grandchild?

B11   What is the value of *x*?

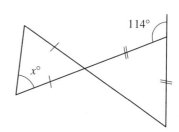

B12   The diagram shows a solid shape.
Each edge is 2 cm long.
What is the surface area of the shape?

B13 A local market stall selling CDs, books and DVDs recorded their sales on a pie chart. Each customer purchased one item.

The table shows the number of each item sold and the angle of the corresponding sector of the pie chart.

Article	Customers	Degrees
CDs	12	90
Books	16	
DVDs		150

What are the missing numbers?

B14 A toy brick is a prism with a right-angled triangular face that has width 5.4 cm and height 3.5 cm. The volume is 37.8 cm^3.

What is the length of the brick?

B15 A worm 3 cm long travels through a wormhole 2 metres long at a steady speed of 84 metres per hour.

How many seconds pass between its 'nose' entering the wormhole and the tip of its 'tail' emerging?

Solutions from the Regional Finals

Group Round Answers

1. 20	6. 3
2. (a) 25; (b) 71	7. 10%
3. (a) HOW ARE YOU (b) VERY WELL THANKS	8. 52
4. 90°	9. (a) 2:1; (b) 3:8
5. 9	10. 25

Crossnumber

	1:3		2:1	3:3	3	4:1		5:1
6:8	1	7:2		6		8:4	0	7
3		9:1	2		10:1	4		2
11:2	12:8	6		13:3	5		14:3	9
	7		15:5		4		7	
16:2	5		17:1	3		18:1	0	19:3
6		20:3	4		21:6	3		5
22:5	2	0		23:5		24:2	25:6	8
7		26:1	3	2	6		4	

Shuttle

A1	8
A2	576
A3	24
A4	360

B1	6
B2	12
B3	50
B4	3750

C1	48
C2	54
C3	15
C4	6

D1	9
D2	2
D3	120
D4	67

Relay

A1	$\frac{3}{10}$	B1	120
A2	350	B2	60
A3	1.65	B3	288
A4	3	B4	50
A5	11	B5	4
A6	32.2 or $32\frac{1}{5}$ or $\frac{161}{5}$	B6	mogs, mags, migs
A7	64	B7	4
A8	110	B8	19
A9	zogs, zags, zigs	B9	3
A10	48	B10	225
A11	3.4	B11	$(4\frac{1}{2},9)$ or $(4.5,9)$ or $(\frac{9}{2},9)$
A12	294	B12	315
A13	37.5 or $37\frac{1}{2}$ or $\frac{75}{2}$	B13	$\frac{13}{60}$
A14	27.04	B14	336
A15	$(1, -1)$	B15	52.8

Solutions from the National Final

Group Circus

1. 251

2. (a) 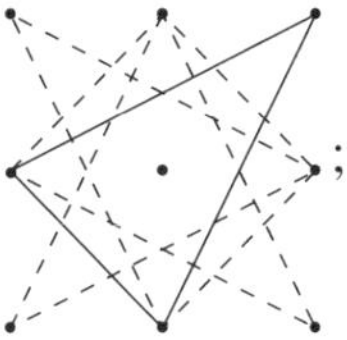 ; (b) 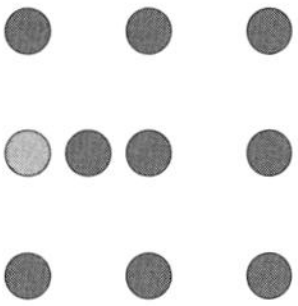

There are 12 distinct lines in a correct solution.

There are other solutions as well as that shown.

3. 5102

4. 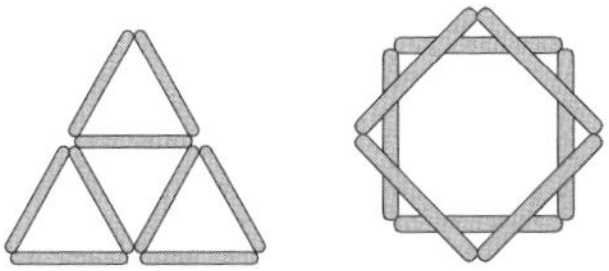

5. (a) 140°, 160°; (b) 144°, 156°; (a) 108°, 162°
The cases may be given in any order. In each case, both angles need to be correct.

6. 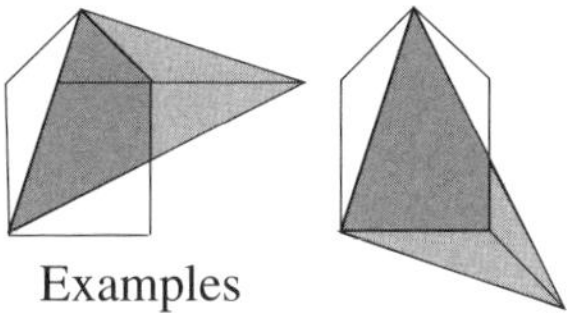

Examples

7. 3 999 960

8. Three equal side totals (17, 19, 20, 21 and 23 are possible).

Crossnumber

(1) 1	1	(2) 1				(3) 5	5	(4) 5
4		(5) 2	3	(6) 4	2	4		9
(7) 7	(8) 2	9		3		(9) 3	7	7
	8		(10) 2	3	(11) 1	0		9
(12) 4	0		8		1		(13) 3	0
1		(14) 3	8	(15) 7	0		2	
(16) 2	3	1		5		(17) 1	4	(18) 3
1		(19) 2	8	4	8	2		4
(20) 8	1	0				(21) 1	3	2

Shuttle

A1	11
A2	8
A3	4
A4	24

B1	21
B2	8
B3	12
B4	8

C1	20
C2	8
C3	6
C4	12.5 or $12\frac{1}{2}$ or $\frac{25}{2}$

D1	12
D2	18
D3	36
D4	16

Relay

A1	20	B1	15
A2	158	B2	4
A3	0.75 or $\frac{3}{4}$	B3	(3, 7)
A4	12	B4	45.40
A5	42	B5	193
A6	(2, 0)	B6	$\frac{111}{10}, \frac{100}{9}, \frac{89}{8}, \frac{78}{7}$
A7	35	B7	23
A8	42	B8	20
A9	7 : 2	B9	80
A10	95	B10	15
A11	161	B11	61.5 or $61\frac{1}{2}$
A12	$\frac{71}{9}, \frac{79}{10}, \frac{87}{11}, \frac{95}{12}$	B12	88
A13	4950	B13	20, 120 (or 120, 20)
A14	6	B14	4
A15	40	B15	87

UKMT and Further Maths Support Programme
Senior Team Maths Challenge 2015

The Senior Team Maths Challenge is now entering into its 9th year and continues to grow in size and popularity. The 2014-15 competition comprised of over 1169 schools competing in 59 Regional Finals held across the United Kingdom; a higher number of competing teams than ever before.

Each team is made up of four students from years 11, 12 and 13 (with a maximum of two Year 13 students per team) and the Regional Competition consists of three rounds; The Group Round, the Crossnumber and the Shuttle. For the Group Round, 10 questions are answered by each team in 40 minutes, while the Crossnumber involves each team solving a mathematical version of a crossword by splitting in two to work on the 'Across' and the 'Down' clues. The competition finishes with the Shuttle, which consists of sets of 4 linked questions, answered in pairs against a timer.

National Final

The culmination of the competition at the National Final in February, where the top 80 teams were invited to the Royal Horticultural Halls in London to compete for the title of 'National Champions', was a wonderful finale, at which the high level of energy and enthusiasm throughout the day created a wonderful celebration of mathematics.

Congratulations to all the schools who took part in the STMC National Final. These schools were:

Adams Grammar School	City of London School
Altrincham Girls' Grammar School	Clifton College
Anthony Gell School	Cockermouth School
Ashford School	Concord College
Atlantic College	Dauntseys School
Bellerbys College, Brighton	Devonport High School for Boys
Bideford College	Dr Challoners Grammar School
Bilborough VI Form College	Dulwich College
Blundell's School	Dunblane High School
Bourne Grammar School	Durham Johnston
Bristol Grammar School	Dyffryn Taf School
Charterhouse School	Eltham College

Ermysteds Grammar School

Eton College

Friends' School Lisburn

George Watson's College

Guernsey Grammar School

Haberdashers' Aske's School for Boys

Hampton School

Harrow School

Hartlepool Sixth Form College

Heaton Manor School

Heckmondwike Grammar School

Hills Road Sixth Form College

King Edward VI Camp Hill Boys' School

King Edward VI Five Ways School

King Edward VI School, Southampton

King Edward's School, Birmingham

King's College London Mathematics School

Kirkham Grammar School

Lancing College

Leventhorpe School

Lincoln Minster School

Loughborough Grammar School

Magdalen College School

Merchiston Castle School

Newcastle Under Lyme School

North London Collegiate School

Norwich School

Oakham School

Oundle School

Parmiter's School

Queen Margaret's School

Rainham Mark Grammar School

Reading School

Rendcomb College

Robert Gordon's College

Robert Smyth Academy

Rossall School

Royal Grammar School Worcester

Royal Grammar School, Newcastle

Ruthin School

S E Essex VI Form College

St Josephs College

St Olave's Grammar School

The Cherwell School

The Crossley Heath School

The Grammar School at Leeds

The King's School, Grantham

The Perse School

The Portsmouth Grammar School

Tiffin School

Torquay Boys' Grammar School

Warwick School

Wellington College

West Kirby Grammar School

Westminster School

Wilson's School

For 2014-15 we had a three-way tie for first place from Hampton School, Harrow School and King Edward's School, Birmingham. The Poster Competition winners were Dunblane High School.

The National Final consisted of the Group Round, the Crossnumber and the Shuttle with the addition of a Poster Competition at the start of the day. Teams were required to answer questions on 'The mathematics of the solar system' and set these in the form of an attractive poster. Thanks to Peter Neumann, Colin Campbell, Karl Hayward-Bradley, Alexandra Hewitt, Kevin Lord and Andrew Jobbings for their hard work in preparing the materials and judging the posters. The Poster Competition did not contribute to the overall result of the National Final but a poster based on the work of the winning team has been professionally produced and printed. This will be sent to all of the schools that took part in the competition.

Thanks

As with all UKMT competitions, thanks must be given to all of the volunteers who wrote questions, acted as checkers for the materials produced, ran Regional Finals alongside FMSP coordinators and who helped on the day at the National Final.

The checkers of the questions were: John Silvester, Alan Slomson, Jenny Ramsden and Martin Perkins.

The 4 Round Rulers, who oversaw the materials for each round, were: Karen Fogden (Group round), Peter Hall (Crossnumber), Mark Harwood (Shuttle) and James Cranch (Starter questions).

The writers of the questions were: Kerry Burnham, Tony Cheslett, Anthony Collieu, David Crawford, Andrew Ginty, James Munro, Charlie Oakley, Dennis Pinshon, Alexandra Randolph and Katie Ray.

As ever, many thanks to everyone involved for making 2014-15 another successful year.

Regional Group Round

1

> When n is a positive integer, $n!$ is the product of all the integers from 1 to n. ($n!$ is read 'n factorial'.)
>
> For example, $5! = 1 \times 2 \times 3 \times 4 \times 5 = 120$.

The expression

$$\frac{(7! + 8! + 9!)^2}{(4! + 5! + 6!)^2}$$

can be written in the form $2^a \times 3^b \times 5^c \times 7^d$.

What is the value of $a + b + c + d$?

2

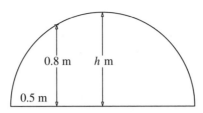

The diagram shows a cross-section of a child's tent. The tent takes the shape of half a cylinder and lies on horizontal ground. At a distance of 0.5 m from the edge of the tent, the height of the tent is 0.8 m above the ground,

The maximum height of the tent is h m. What is the value of h?

3   Three tangents are drawn to a circle at points A, B and C.

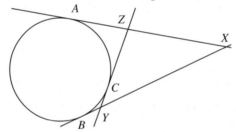

The tangents intersect at points X, Y and Z as shown.

The length of BY is 1 cm and the length of YX is 4 cm.

What is the length of the perimeter of the triangle XYZ, in cm?

4   A fashion designer wants to print a flower motif onto some material. The design consists of a circular centre and four petals. The edges of the petals are congruent semi-circular arcs AB, BC, CD and DA.

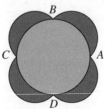

Calculate the ratio of the total area of the petals to the area of the circle.

Give your answer in the form $1 : x$.

5 Xana and Zara start together in a corner of a large square field of side 40 m. They set off at the same time and each walks around the edge of the field, one clockwise and the other anti-clockwise.

The ratio of Xana's speed to Zara's speed is 3 : 5. When they meet each other for the first time, and on each subsequent meeting, they reverse their directions and swap their speeds.

How far from their starting position are they when the meet each other for the 200th time, in metres?

6 Each of the ten different letters in *MATHS CHALLENGE* represents a different digit.

$$\begin{array}{cccccc} & S & T & M & C \\ + & S & T & M & C \\ \hline M & A & T & H & S \end{array}$$

$L^2 = E$ and $E > G$.

What is the value of $C + H + A + L + L + E + N + G + E$?

7 Three of the vertices of the parallelogram *PQRS* lie on the coordinate axes at *P*, *Q* and *R* as shown.

The graph of

$$y = x^2 - x - 12$$

passes through *P*, *Q* and *R*.

What is the area of the parallelogram *PQRS*?

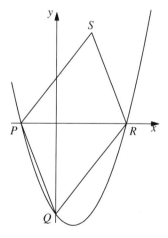

8   A sequence of non-negative integers with nth term u_n is defined by $u_1 = a$, $u_2 = b$ and $u_{n+2} = u_{n+1} + u_n$ for all positive integers n.

How many pairs of non-negative integers (a, b) are there such that all of the following are true?

    (i)   21 is a term of the sequence.

    (ii) $a \neq 21$.

    (iii) $b \neq 21$.

9

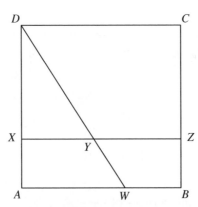

Square $ABCD$ has side length 2 units. The square is cut along the straight lines DW and XZ, producing three similar trapeziums and a right-angled triangle DXY.

What is the area of the triangle DXY?

Give your answer in the form $a\sqrt{5} + b$ where a and b are integers.

10   A cuboid has faces with areas $6\,\text{cm}^2$, $9\,\text{cm}^2$ and $24\,\text{cm}^2$.

What is the volume of the cuboid, in cubic centimetres?

Regional Final Crossnumber

Across

1   The value of $a + b$ when $(2 + \sqrt{2})(3 + \sqrt{8})$ is written in the form $a + b\sqrt{2}$. (2)

3   One quarter of the product of the square of 20 Down and the cube root of 18 Down (3)

6   The value of $(23 \text{ Across})^2 \times (20 \text{ Down})^{-1}$ (3)

7   The sum of 9 Down and 21 Across (3)

9   The positive solution of $\dfrac{x}{2} + \dfrac{2}{x + 1} = \dfrac{17}{3}$ (2)

10   A prime number (2)

12   The value of y when x is 6 Across given that y is proportional to x and y is 6 Across when x is 23 Across (3)

14   The sum of the two prime numbers nearest to 23 Across (3)

16   The prime number between 10 and 20 that is not a solution to any other clues (2)

18   The mean of 1 Across, 5 Down and 18 Across (2)

19   The square of a prime number, not divisible by 31 (3)

21   One more than 9 Down (3)

22   Half of the sum of the interior angles in a polygon with 9 Across sides (3)

23   The coefficient of x^2 in the expansion of $(2 + 3x)^3$ (2)

Down

1	A multiple of three	(3)
3	The value of 2×8 Down -12 Down $+2$	(3)
4	A multiple of 5 Down	(2)
5	The positive solution of $\dfrac{20}{x-3} + \dfrac{32}{x+3} = 4$	(2)
8	A number which makes the sum of the digits in the seventh column a multiple of seven	(3)
9	One less than 21 Across	(3)
11	The value of 6 Across $-$ 14 Across $-$ 20 Down	(2)
12	The square of a prime number	(2)
13	The value of $(18 \text{ Down})^{4/3}$	(3)
15	The value of $24y$ when x is 23 Across, given that y is inversely proportional to x and y is 1 Across when x is 20 Down	(3)
17	Ten times the product of two prime numbers	(3)
18	A cube	(3)
19	A square	(2)
20	The value of $a + b$ when $(3 + \sqrt{2})(2 + \sqrt{8})$ is written in the form $a + b\sqrt{2}$	(2)

Regional Shuttle

A1 Find the units digit of 83^3.

Pass on *three less* than your answer.

A2 *T is the number that you will receive.*

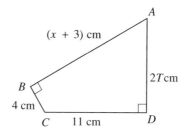

Pass on the value of *x*.

A3 *T is the number that you will receive.*

In the diagram below, lines *AB* and *EF* are parallel.

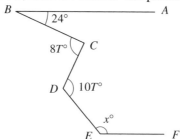

Pass on the sum of the digits of *x*.

A4 *T is the number that you will receive.*

In the diagram, the equation of line *CB* is $y = x + T$ and the *x* coordinate of *B* is 6.

The area of the quadrilateral *ABCO* is 110 square units.

Write down the *x*-coordinate of the point *A*.

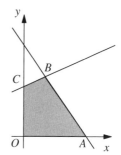

B1   10% of 20% of 30% of 40% of £50 $=$ x pence.

Pass on the value of x.

B2   *T is the number that you will receive.*

Express

$$\frac{\sqrt{63} \times \sqrt{13T}}{2\sqrt{91} \times \sqrt{30}}$$

in the form

$$\sqrt{\frac{a}{b}},$$

where a and b are positive integers with no common factor other than 1.

Pass on the value of $b - a + 2$.

B3   *T is the number that you will receive.*

Reflect the point $\left(\frac{3}{2}, T - 4\right)$ in the line $x = -\frac{9}{4}$.

Reflect the new point in the line $y = 5$.

The final point is now (a, b).

Pass on *two less than* the value of $a + b$.

B4   *T is the number that you will receive.*

Evaluate

$$\frac{1 - p^2}{p + p^2}$$

where $p = \dfrac{1}{T + 2}$.

Write down your answer.

C1   The diagram shows a quadrilateral drawn within a circle with centre O.

Vertices A, B and C lie on the circumference of the circle.

Pass on the value of x.

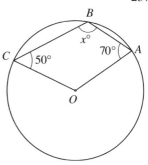

C2   *T is the number that you will receive.*

The surface area of a cylindrical flask is 64 cm^2 and its volume is T cm^3.

A similar flask has surface area of 400 cm^2 and volume of V cm^3.

Pass on the sum of the digits of V.

C3   *T is the number that you will receive.*

x and y satisfy the simultaneous equations

$$5x - 2y = 0$$

$$2xy = 5T - 20x.$$

Pass on the positive value of x.

C4   *T is the number that you will receive.*

A bag contains x red counters and T blue counters.

Two counters are removed from the bag and are not replaced.

The probability of obtaining two counters of different colours is 0.5.

Write down the larger of the possible values of x.

D1   Solve the equation

$$3^{x^2} \times 9^{2x} \times 27 = 1.$$

The solutions are $x = A$ and $x = B$.

Pass on the value of $3AB - 1$.

[*Hint: convert all components to powers of* 3.]

238

D2   *T is the number that you will receive.*

The equation of the perpendicular bisector of the line joining $\left(\frac{1}{4}T,6\right)$ and $(T + 2,10)$ can be expressed in the form $y = Ax + B$. Pass on the value of $\frac{1}{6}(B + A)$.

D3   *T is the number that you will receive.*

When the following expression is expanded, the coefficients of x^3 and x are A and B respectively.

$$(x - 3)(x + T - 1)(2x - 1)$$

Pass on the value of $A - B - 5$.

D4   *T is the number that you will receive.*

Calculate the value of

$$(2T)^{3/2} \times 8^{-2/3} \times 2^{-1}.$$

Write down your answer.

Group Round answers

1.	Value of $a + b + c + d$	8
2.	Value of h	0.89
3.	Length of perimeter of triangle XYZ	10
4.	Total area of petals : area of circle	$1 : \frac{\pi}{2}$
5.	Distance from starting position	0
6.	Value of $C + H + A + L + L + E + N + G + E$	48
7.	Area of parallelogram $PQRS$	84
8.	Number of pairs (a, b)	40
9.	Area of triangle DXY	$2\sqrt{5} - 4$
10.	Volume of cuboid	36

Crossnumber: Completed grid

[1] 1	[2] 7		[3] 4	0	[4] 5		[5] 1
	[6] 1	6	2		[7] 2	[8] 2	3
[9] 1	1		[10] 9	7		3	
1		[11] 3			[12] 4	8	[13] 6
[14] 1	[15] 1	2			9		2
	3		[16] 1	[17] 9		[18] 1	5
[19] 3	6	[20] 1		[21] 1	1	2	
6		[22] 8	1	0		[23] 5	4

Shuttle answers

A1	4
A2	10
A3	10
A4	10

B1	12
B2	3
B3	3
B4	4

C1	120
C2	21
C3	3
C4	6

D1	8
D2	3
D3	8
D4	8

National Final Group Round

1   When n is a positive integer, $n!$, read 'n factorial', is the product of all the integers from 1 to n.
For example, $5! = 1 \times 2 \times 3 \times 4 \times 5$.

What is the highest power of 6 that is a factor of $66!$?

2   Five pirates agree to split a number of coins in the following way.
The first pirate takes half the coins and one more.
The second pirate takes a third of the remainder and two more.
The third pirate takes a quarter of the remaining coins and three more.
The fourth pirate takes a fifth of the remaining coins and four more.
The fifth pirate takes what's left.

Given that each pirate receives a whole number of coins and the fifth pirate receives less than the third, how many coins did the first pirate take?

3   In how many different ways can 2015 be written as the difference of the squares of two positive integers?

4   *OPQA* is a rectangle with *OP* = 3 units and *PQ* = 2 units.

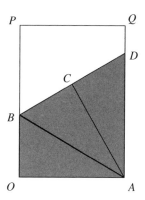

Three congruent triangles *AOB*, *ACB* and *ACD* are removed from the rectangle.
What is the area of the remaining trapezium?

5   At Mary Box School every pupil in Year 10 studies at least two sciences.
There are 220 students in the year group, of which 150 study biology, 180 study chemistry and 170 study physics.
How many students study all three sciences?

6   In the right-angled triangle shown the sides adjacent to the right angle have lengths 7 and 24.

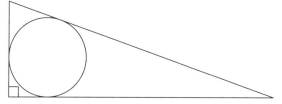

A circle is inscribed in the triangle.
What is the area of the circle as a multiple of π?

7   A sequence is defined by the recurrence relation

$$u_n = (u_{n-1})^2 - (u_{n-2})^2 \qquad \text{and}$$

$$u_1 = u_2 = 1.$$

What is the value of u_{100}?

8   2015 is the 400th anniversary of the birth of Frans van Schooten. His significant contribution to mathematics was to take the new ideas of René Descartes' cartesian geometry and spread them in an understandable way. He also suggested that coordinate geometry could be extended to 3-dimensional space.

The equation of a sphere, centred at the origin, can be written as $x^2 + y^2 + z^2 = r^2$, where r is the radius of the sphere.

What is the volume of the largest cuboid with sides of integer length that can fit inside a sphere of radius 9 units?

9   In a hexagon *ABCDEF* the four interior angles *ABC*, *BCD*, *CDE* and *DEF* are each equal to 135°.

Also, $AB = 1$, $BC = 2$, $CD = 4$, $DE = 8$, $EF = 16$ and $FA = x$.

What is the value of x^2?

10   In a street there are five houses, each painted a different colour.

A person of a different nationality lives in each house.

Each of these five people drinks a different type of beverage, plays a different type of sport and keeps a different type of pet.

The Brit lives in the red house.

The Swede keeps dogs as pets.

The Dane drinks tea.

The green house is next to, and on the left of the white house.

The owner of the green house drinks coffee.

The person who plays hockey rears birds.

The owner of the yellow house goes swimming.

The person living in the centre house drinks milk.

The Norwegian lives in the first house.

The person who plays football lives next to the one who keeps cats.

The person who keeps horses lives next to the person who goes swimming.

The person who cycles drinks juice.

The Italian plays tennis.

The Norwegian lives next to the blue house.

The person who plays football has a neighbour who drinks water.

What is the nationality of the person who owns the fish?

National Final Crossnumber

Across

1. A value of a where $\sqrt{4949}$ is written as $k\sqrt{a}$ and k is an integer (3)
4. Larger than a sixth of 9 Across and also twice a prime (3)
6. A power of four (5)
7. The value of $n^2(n - 1) + 2n$ where $n = \dfrac{10\text{ Down}}{8\text{ Down}}$ (3)
9. A multiple of 61 (3)
11. The coefficient of x^2 in the expansion of $(6 + x)^4$ multiplied by five more than the square root of 5 Down (4)
12. $(4\text{ Down})^{3/2} + (6\text{ Across})^{1/4}$ (4)
14. The sixth root of 8 Down multiplied by 111 (3)
16. This makes the mean of 16 Across, 16 Down, 17 Down and 20 Across one less than a square (3)
18. A quarter of 6 Across (5)
19. A square (3)
20. One more than the mean of 14 Across, 15 Down, 14 Down and 19 Across (3)

Down

1. The value of $20 + 3^a$ where $a = \dfrac{14 \text{ Across} - 111}{111}$ (3)

2. Five times a fifth power (3)

3. The coefficient of x in the expansion of $(5 + bx)^3$ where b is equal to 1 Across (4)

4. A square (3)

5. The value of $n\left(n^2 - n + 1\right)^2$ where $n = \dfrac{6 \text{ Across}}{18 \text{ Across}}$ (3)

8. A fifth of 10 Down (5)

10. A power of 5 (5)

13. Half of the product of x and y given by the simultaneous equations $3x + 4y = 754$; $4x + 5y = 970$ (4)

14. A fifteenth of 3 Down (3)

15. A multiple of 43 (3)

16. This makes the mean of 16 Across, 16 Down, 17 Down and 20 Across one less than a square (3)

17. One less than $(a - 1)(b - 1)$ where $a = (8 \text{ Down})^{1/3}$ and $b = \left(\tfrac{1}{4} \text{ of 18 Across}\right)^{1/3}$ (3)

Shuttle

A1   Evaluate

$$\frac{7! \times 3!}{4!} \times \frac{5!}{6! \times 3!}.$$

Pass on *five less* than your answer.

A2   *T is the number that you will receive.*

Integers m, n, p and q are such that

$$2^m + 2^n + 2^p + 2^q = T.$$

Pass on the value of $m + n + p + q$.

A3   *T is the number that you will receive.*

Alan and Bob drive separately along the same $4T$ miles of a motorway.

Alan drives at a constant speed of 50 mph.

Bob drives the first $2T$ miles at 60 mph and the remaining $2T$ miles at 40 mph.

Find the difference. in seconds, between their journey times.

Pass on your answer.

A4   *T is the number that you will receive.*

$$P + S = C$$

where P, S and C are a prime, a square and a cube respectively. They are all less than T.

Write the number of different solutions.

B1   Each of the two different solutions to the equation

$$4x^2 + ax + b = 0$$

is 3 less than one of the two different solutions to the equation

$$4x^2 - 12x + 5 = 0.$$

Pass on the value of b.

246

B2   *T is the number that you will receive.*

The mean of a set of four positive integers is T.
The median of the same set of integers is $T + 1$.
Pass on the largest value of the range.

B3   *T is the number that you will receive.*

$$x \; = \; \frac{16^{3/4} \times 8^{-2/3} \times 4^{1/2}}{(T + 2)^{-5/3}}.$$

Pass on the sum of the digits of x.

B4   *T is the number that you will receive.*

The diagram shows a
sector of a circle with
radius $(T - 1)$ cm and
angle $AOB = 216°$.

The sector is cut out of
paper and bent until A and
B meet to form a cone. The
volume of the cone is
$k\pi$ cm^3.

Write down the value of k.

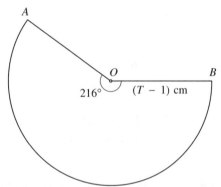

C1   Find the equation of the perpendicular bisector of the line segment
joining the points $(10, \; 4)$ and $(8, -2)$.
Your answer can be expressed in the form $by \; = \; ax + c$ where a,
b and c are integers.

Pass on the value of $\dfrac{c}{b}$,

C2   *T is the number that you will receive.*

Solve the equation

$$9^{2x} \; = \; \left(\frac{1}{3}\right)^{T + 1}.$$

Your answer can be expressed in the form $-\dfrac{a}{b}$, where a and b are
positive integers with no common factor greater than 1.
Pass on the value of $a - b$.

C3 *T is the number that you will receive.*

An irregular pentagon has angles $(150 + 20T)°$, $50T°$, $4x°$, $3x°$ and $x°$.

Pass on the value of x.

C4 *T is the number that you will receive.*

The diagram shows a circle with centre O.

Angle $AOD = (T + 20)°$.

Length $BC = 5$ cm.

Write down the area of triangle BCD.

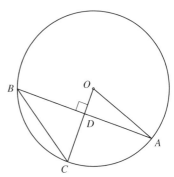

D1 Evaluate

$$1 + \cfrac{1}{1 + \cfrac{1}{1 + 2!}}.$$

Your answer can be expressed as a fully simplified fraction $\dfrac{a}{b}$, where a and b are positive integers.

Pass on the value of $a + b$.

D2 *T is the number that you will receive.*

The angles in a hexagon form an arithmetic sequence with common difference $T + 9$ degrees.

The size of the smallest angle is $A°$.

Pass on the value of $A - 10$.

248

D3   *T is the number that you will receive.*

Simplify

$$\frac{9x^2 + 21x + 6}{3x^2 - 8x - 3} \times \frac{4x^2 - 10x - \frac{1}{10}T}{2x^3 + 9x^2 + 12x + 4}.$$

Your answer can be expressed in the form $\dfrac{a}{x + b}$.

Pass on the value of $a - b$.

D4   *T is the number that you will receive.*

The ratio of the surface areas of two solid hemispheres is $1 : 4$.
The radius of the larger hemisphere is $\dfrac{5T}{2}$ cm.

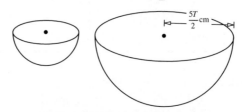

Write down the positive difference between the two volumes.

Group Round answers

1.	31	6.	9π
2.	41	7.	-1
3.	4	8.	1100
4.	$6 - 2\sqrt{3}$	9.	$309 + 130\sqrt{2}$
5.	60	10.	Italian

Crossnumber: Completed grid

¹1	0	²1	■	³7	■	⁴1	6	⁵6
0	■	⁶6	5	5	3	6	■	7
⁷1	⁸1	0	■	7	■	⁹9	¹⁰7	6
■	5	■	■	5	■	■	8	■
¹¹6	6	9	6	■	¹²2	2	1	3
■	2	■	¹³5	■	■	■	2	■
¹⁴5	5	¹⁵5	■	8	■	¹⁶3	5	¹⁷3
0	■	¹⁸1	6	3	8	4	■	5
¹⁹5	7	6	■	0	■	²⁰5	3	9

Shuttle answers

A1	30
A2	10
A3	120
A4	4

B1	5
B2	6
B3	11
B4	96

C1	4
C2	1
C3	40
C4	$\frac{25\sqrt{3}}{8}$

D1	11
D2	60
D3	4
D4	$\frac{1750}{3}\pi$

Other aspects of the UKMT

As well as the Maths Challenges, the UKMT is involved in other events and activities.

Enriching Mathematical Thinking
UKMT Teacher Meetings 2015

Six teacher meetings were held this year: Aberdeen (University of Aberdeen), Cambridge (University of Cambridge), Coventry (Coventry University), Greenwich (University of Greenwich), Oxford (University of Oxford) and York (National STEM Centre, University of York)

Around 285 teachers attended the one-day events. Each meeting featured three sessions with lunch and refreshment breaks and delegates received a resource pack and CPD certificate to take back to the classroom.

NRICH (www.nrich.maths.org.uk) gave sessions at all six meetings and we are grateful to Charlie Gilderdale, Alison Kiddle and Frances Watson for the quality of these sessions and the accompanying resources.

Rob Eastaway, author and Director of Maths Inspiration gave an inspiring talk on the subject of Mathematical Modelling at Coventry; Katie Chicot engaged delegates at Oxford with her session on Maths Mindsets; James Grime enthused our Cambridge teachers with a talk on Coding and the Enigma Machine; and Colin Wright captured the imagination of the audience with his display on the Mathematics of Juggling at Aberdeen, and a session on Mobius Strips at Greenwich.

UKMT volunteers led a session at each event demonstrating the mathematical thinking behind the questions used in the UK Maths Challenges and the Team Maths Challenges, and how UKMT materials can be used to stimulate classroom interest. The 2015 speakers were Vicky Neale, Steven O'Hagan, Steve Mulligan, Stephen Power and Dominic Rowland.

We are also very grateful to our host venues and to all UKMT volunteers who assisted with these events.

Mathematical Circles

The Mathematical Circles developed from two trial events in spring 2012. Following on from the success of these, a further two events were run in early 2013 in Glasgow and Leeds. Thanks to a two-year grant from the Department for Education, we were able to expand these events from April 2013.

Local schools are invited to select two students from Year 10 (and equivalent) to send to the two-day events which are comprised of mathematically demanding work through topics such as geometry, proof and modular arithmetic. Students have the opportunity to discuss mathematics and make new friends from other schools around their region.

Our thanks go to the following people who ran the events, and to the schools who supported these:

Winchester, St Swithun's School, run by Stephen Power

Wells, Wells Cathedral School, run by Susie Jameson

Oldham, North Chadderton School, run by Alan Slomson

Harrogate, Harrogate Ladies College, run by James Welham

Lincoln, Lincoln Minster School, run by John Slater

Liverpool, Carmel College, run by Dean Bunnell

London West, St Pauls School, run by Dominic Rowland

Teesside, Conyers School, run by Anne Baker

Oundle, Oundle School, run by Alan Slomson

Glasgow, Hutchesons' Grammar School, run by Catherine Ramsay

Gloucester, Wycliffe School, run by Mark Dennis

London West, St Pauls School, run by Dominic Rowland (Year 12 trial event)

A sample timetable from Lincoln is given below:

	Day 1	Day 2
9.00	Arrival and registration	Arrival and registration
09.45	Warming up problems and getting to know you	Combinatorics
10.30	Break	Break
10.45	Spirals and loops	Finite arithmetic
12.00	Lunch	Lunch
12.45	Discrete processes	What Boole started
14.00	Break	Break
14.15	How to drill a square hole	Big numbers
15.30	Close	Close

Thanks are also given to those people who ran sessions at these events (a list of which is given at in the Volunteers section of the Yearbook).

Mathematical Circles are being run throughout the next academic year. If you would like to find out more about how you can become involved in the Mathematical Circles, either through your school hosting an event or by supporting us in running a session, please do contact us at enquiry@ukmt.org.uk.

Primary Team Maths Resources

In recognition and support of the growing number of secondary schools organising and hosting local team maths events for their feeder schools, UKMT developed a set of Primary Team Maths Resources (PTMR) intended for use at such events. The first ever PTMR were made available free of charge in spring 2012. At the start of each calendar year, a new set of material has since been made available.

Schools may choose to use the materials in other ways, e.g. a primary school may use the materials to run a competition for their own Year 5 and 6 pupils (and equivalent), or a secondary school may use the materials as an end of term activity for their Year 7 pupils.

The PTMR include more materials than would be needed for any one competition, allowing schools to pick and choose those most appropriate for their purposes. Some of the rounds are familiar from the UKMT Team Challenges (the Group Round, Crossnumber, Relay and Mini Relay) and the material included some new rounds (the Logic Round, Make a Number, Open Ended Questions, and Speed Test).

The 2015 PTMR and full instructions for suggested use is available by contacting the UKMT via email at enquiry@ukmt.org.uk. Further details including sample and past materials can be found on our website at http://www.ukmt.org.uk/team-challenges/primary-team-maths-resources/ .

Best in School Events

Since 2007, the UKMT has partnered with the Royal Institution to invite top scoring IMC candidates to attend RI masterclass celebration events, held in the summer term. These events give students from Year 9 (and equivalent) and sometimes Year 12 (and equivalent) the opportunity to attend inspiring lectures, meet mathematicians from their local area, and have a go at 'hands-on' mathematics.

In 2015, these events took place in Edinburgh, Liverpool, London (Year 9 and Year 12 events), Newcastle, and Plymouth.

Website – www.ukmt.org.uk

Visit the UKMT's website for information about all the UKMT's activities, including the Maths Challenges, team events, latest UKMT news and newsletters, contact details, and to purchase publications and past papers.

There are online resources featuring past questions from the Challenges, mentoring questions, and sample Primary Team Maths Challenge materials. Also links to sponsors, supporters and other mathematical bodies providing further resources for young mathematicians.

Other similar bodies overseas

The UKMT has links of varying degrees of formality with several similar organisations in other countries. It is also a member of the World Federation of National Mathematics Competitions (WFNMC). What follows is a brief description of some of these other organisations. Some of the information is taken from the organisations' web sites but a UK slant has been applied.

"Association Kangourou sans Frontières"

http://www.aksf.org/

The obvious question is: why Kangaroo? The name was given in tribute to the pioneering efforts of the Australian Mathematics Trust. The Kangaroo contest is run by local organisers in each country under the auspices of the 'Association Kangourou sans Frontières', which was founded by a small group of countries in 1991. There are now over 50 countries involved and more than six million participants throughout Europe and beyond, from the UK to Mongolia and from Norway to Cyprus.

In the UK in 2015, over 7000 children in the years equivalent to English Years 9, 10 and 11 took part in the 'Cadet' and 'Junior' levels of the Kangaroo competition, as a follow-up to the Intermediate Maths Challenge. Three representatives of the UK Mathematics Trust, Andrew Jobbings, Paul Murray and David Crawford, attended the meeting in Puerto Rico, at which these Kangaroo papers were constructed.

The main objective of the Kangaroo, like all the competitions described in this section, is to stimulate and motivate large numbers of pupils, as well as to contribute to the development of a mathematical culture which will be accessible to, and enjoyed by, many children and young people. The Association also encourages cross-cultural activities; in some countries, for example, prize-winners are invited to attend a mathematics 'camp' with similar participants from other nations.

The Australian Mathematics Trust

www.amt.edu.au

For over twenty-five years, the Australian Mathematics Competition has been one of the major events on the Australian Education Calendar, with about one in three Australian secondary students entering each year to test their skills. That's over half a million participants a year.

The Competition commenced in 1978 under the leadership of the late Professor Peter O'Halloran, of the University of Canberra, after a successful pilot scheme had run in Canberra for two years.

The questions are multiple-choice and students have 75 minutes in which to answer 30 questions. There are follow-up rounds for high scorers.

In common with the other organisations described here, the AMC also extends its mathematical enrichment activities by publishing high quality material which can be used in the classroom.

Whilst the AMC provides students all over Australia with an opportunity to solve the same problems on the same day, it is also an international event, with most of the countries of the Pacific and South-East Asia participating, as well as a few schools from further afield. New Zealand and Singapore each enter a further 30,000 students to help give the Competition an international flavour.

World Federation of National Mathematics Competitions – WFNMC

www.amt.canberra.edu.au/wfnmc.html

The Federation was created in 1984 during the Fifth International Congress for Mathematical Education.

The Federation aims to provide a focal point for those interested in, and concerned with, conducting national mathematics competitions for the purpose of stimulating the learning of mathematics. Its objectives include:

- Serving as a resource for the exchange of information and ideas on mathematics competitions through publications and conferences.
- Assisting with the development and improvement of mathematics competitions.

- Increasing public awareness of the role of mathematics competitions in the education of all students and ensuring that the importance of that role is properly recognised in academic circles.
- Creating and enhancing professional links between mathematicians involved in competitions around the world.

The World Federation of National Mathematics Competitions is an organisation of national mathematics competitions affiliated as a Special Interest Group of the International Commission for Mathematical Instruction (ICMI).

It administers a number of activities, including

- The Journal *Mathematics Competitions*
- An international conference every four years.
- David Hilbert and Paul Erdős Awards for mathematicians prominent on an international or national scale in mathematical enrichment activities.

The UKMT sent two delegates, Tony Gardiner and Bill Richardson, to the WFNMC conference in Zhong Shan in 1998 and provided support for several delegates who attended ICME 9 in Tokyo in August 2000, at which the WFNMC provided a strand.

In August 2002, the WFNMC held another conference, similar to the one in 1998. The venue for this was Melbourne, Victoria. On this occasion, the UKMT provided support for two delegates: Howard Groves and Bill Richardson.

In July 2006, WFNMC 5 was held in the UK at Robinson College, Cambridge. This event was a tremendous success with around 100 delegates from many parts of the world.

In July 2008, WFNMC had a strand at ICME 11 in Mexico. UKMT was represented by Bill Richardson.

In July 2010, WFNMC 6 was held in Riga. The UKMT was represented by Howard Groves, Dean Bunnell, David Crawford and James Welham.

In July 2014, WFNMC 7 was held in Colombia. The UKMT was represented by David Crawford.

Lists of volunteers involved in the UKMT's activities

Members of the BMOS Extended Committee

Robin Bhattacharyya (Loughborough GS)   Philip Coggins (ex Bedford School)
Mary Teresa Fyfe (Hutchesons' GS)   James Gazet (Eton College)
Ben Green (Trinity College, Cambridge)   Andrew Jobbings (Arbelos)
Jeremy King (Tonbridge School)   Patricia King (ex Benenden School, Kent)
Gerry Leversha (ex St Paul's School)   Adam McBride (Uni. of Strathclyde)
David Monk (ex Edinburgh University)   Peter Neumann (Queen's Coll., Oxford)
Alan Pears (ex King's College, London)   Adrian Sanders (ex Trinity College, Camb.)
Zhivko Stoyanov (University of Bath)   Alan West (ex Leeds University)
Brian Wilson (ex Royal Holloway, London)

BMOS Markers

James Aaronson (Trinity Coll, Cambridge)   Ben Barrett (Trinity Coll, Cambridge)
Natalie Behague (Trinity Coll, Cambridge) Alexander Betts (Trinity Coll, Cambridge)
Robin Bhattacharrya (Loughborough GS)   Matija Bucic (Trinity Coll, Cambridge)
Andrew Carlotti (Trinity Coll, Cambridge) Ilya Chevyrev (University of Oxford)
Philip Coggins (ex Bedford School)   James Cranch (University of Sheffield)
Tim Cross (KES, Birmingham)   Ceri Fiddes (Millfield School)
Richard Freeland (Trinity Coll, Cambridge) James Gazet (Eton College)
Gabriel Gendler (Trinity Coll, Cambridge) Ed Godfrey (Trinity Coll, Cambridge)
Daniel Griller (Hampton School)   Jo Harbour (Mayfield Pri. Sch, Cambridge)
Karl Hayward-Bradley (Wellington Coll, Shanghai)
John Haslegrave (University of Sheffield) Tim Hennock (Jane Street Capital)
Ina Hughes (University of Leeds)   Ian Jackson (Tonbridge School)
Andrew Jobbings (Arbelos)   Vesna Kadelburg (Stephen Perse Found.)
Jeremy King (Tonbridge Sch)   Gerry Leversha (ex St Paul's School)
Sam Maltby (New Vision)   Matei Mandache (Trinity Coll, Cambridge)
Freddie Manners (University of Oxford)   David Mestel (Trinity Coll, Cambridge)
Jordan Millar (Trinity Coll, Cambridge)   Anca Mustata (University College, Cork)
Joseph Myers (CodeSourcery, Inc)   Peter Neumann (ex Queen's College, Ox.)
Sylvia Neumann (Oxford)   Craig Newbold (Jane Street Capital)
Martin Orr (UCL)   Jenny Owladi (Bank of England)
Preeyan Parmar (Trinity Coll, Cambridge)   Roger Patterson (Sevenoaks School)
David Phillips (Trinity Coll, Cambridge)   Hannah Roberts (University of Oxford)
Dominic Rowland (St Paul's School)   Paul Russell (Churchill Coll., Cambridge)
Geoff Smith (Uni. of Bath)   Karthik Tadinada (St Paul's School)
Jerome Watson (Bedford School)   Dominic Yeo (University of Oxford)

MOG Markers

Ben Barrett (Trinity Coll, Cambridge)
Lax Chan (Open University)
Philip Coggins (ex Bedford Sch)
Sue Cubbon (St Albans, Herts)
James Gazet (Eton College)
Jo Harbour (Mayfield Pri. Sch., Cambridge)
Ina Hughes (Univ of Leeds)
Magdalena Jasicova (Cambridge)
David Mestel (Trinity Coll, Cambridge)
Vicky Neale (Murray Edwards College)
Sylvia Neumann (Oxford)
Martin Orr (University College London)
David Phillips (Trinity Coll, Cambridge)
Jenni Sambrook (Uckfield)
Jerome Watson (Bedford School)

Andrew Carlotti (Trinity Coll, Cambridge)
Andrea Chlebikova (St Catharine's Cambridge)
Tim Cross (King Edward's Birmingham)
Paul Fannon (The Stephen Perse Found.)
Adam Goucher (Trinity Coll, Cambridge)
Maria Holdcroft (Queen's Coll, Oxford)
Freddie Illingworth (Oxford)
Vesna Kadelburg (Stephen Perse Found.)
Joseph Myers (CodeSourcery, Inc.)
Peter Neumann (ex Queen's Coll, Oxford)
Craig Newbold (Jane Street Capital)
Preeyan Parmar (Trinity Coll, Cambridge)
Linden Ralph (Trinity Coll., Cambridge)
Marius Tiba (Trinity Hall, Cambridge)
Brian Wilson (ex Royal Holloway, London)

Markers for IMOK and JMO

Anne Baker	(Conyers School, Stockton-on-Tees)	IMOK
Natalie Behague	(Trinity College, Cambridge)	IMOK
Dean Bunnell	(ex Queen Elizabeth GS, Wakefield)	IMOK
Valerie Chapman	(Northwich)	IMOK
Mike Clapper	(Australian Mathematics Trust)	JMO
Philip Coggins	(ex Bedford School)	IMOK / JMO
Madeline Copin	(North London Collegiate School)	IMOK
James Cranch	(University of Sheffield)	IMOK / JMO
David Crawford	(Leicester Grammar School)	IMOK / JMO
Tim Cross	(KES, Birmingham)	IMOK
Sue Cubbon	(St Albans, Herts)	IMOK
Wendy Dersley	(Southwold)	JMO
David Forster	(Oratory School)	IMOK
Mary Teresa Fyfe	(Hutchesons' Grammar School, Glasgow)	IMOK / JMO
Carol Gainlall	(Park House School, Newbury)	IMOK / JMO
Gwyn Gardiner	(ex King Edward's School, Birmingham)	JMO
Tony Gardiner	(Birmingham)	JMO
James Gazet	(Eton College)	IMOK
Nick Geere	(Okehampton)	IMOK
Michael Griffiths	(Warrington)	IMOK
Howard Groves	(ex RGS, Worcester)	JMO
James Hall	(Harrow School)	IMOK

Hugh Hill	(Winchester College)	IMOK
Ina Hughes	(University of Leeds)	IMOK
Carl James	(Leicester Grammar School)	IMOK
Andrew Jobbings	(Arbelos, Shipley)	IMOK / JMO
David Knipe	(Cambridge)	IMOK
Aleksandar Lishkov	(Oxford)	IMOK
Nick Lord	(Tonbridge School)	IMOK
Sam Maltby	(Sheffield)	IMOK / JMO
Linda Moon	(The Glasgow Academy)	IMOK
Phil Moon	(The High School of Glasgow)	IMOK
Peter Neumann	(The Queen's College, Oxford)	IMOK / JMO
Sylvia Neumann	(Oxford)	IMOK
Steven O'Hagan	(Hutchesons' Grammar School,Glasgow)	IMOK / JMO
Jenny Perkins	(Torbridge High School, Plymouth)	JMO
Stephen Power	(St Swithuns School, Winchester)	IMOK / JMO
Laurence Rackham	(King Edward's School, Birmingham)	JMO
Catherine Ramsay	(Hutchesons' Grammar School)	JMO
Jenny Ramsden	(High Wycombe)	IMOK
Christine Randall	(Southampton)	IMOK
Peter Ransom	(Southampton)	JMO
Lionel Richard	(Frankfurt International School)	JMO
Chris Robson	(University of Leeds)	IMOK
Fiona Shen	(Queen Ethelburga's College, York)	IMOK / JMO
John Slater	(Market Rasen)	IMOK / JMO
Alan Slomson	(ex University of Leeds)	IMOK / JMO
Jon Stone	(St Paul's School, London)	IMOK
Anne Strong	(Romsey)	IMOK
Karthik Tadinada	(St Paul's School, London)	IMOK
Christopher Walker	(Cumnor House School)	IMOK
Paul Walter	(Highgate School, London)	IMOK
Jerome Watson	(Bedford School)	IMOK
David Webber	(University of Glasgow)	IMOK / JMO
Michaela Weiserova	(Surrey)	IMOK
Brian Wilson	(ex Royal Holloway, London)	IMOK
Rosie Wiltshire	(Wootton Bassett School)	IMOK
Heather Yorston	(University of Strathclyde)	IMOK

Problems Groups

There are currently five groups. The first being the BMO Setting Committee.

Jeremy King	(Chair) (Tonbridge School)
Julian Gilbey	(London)
Paul Jefferys	(ex Trinity College, Cambridge)
Gerry Leversha	(ex St Paul's School)
Jack Shotton	(Imperial College, London)
Geoff Smith	(University of Bath)

The other six groups have overlapping membership. There is one group for each and the chair is shown in []: the Senior Mathematical Challenge (S) [Karen Fogden]; the Junior and Intermediate Mathematical Challenges (I&J) [Howard Groves]; the Junior Mathematical Olympiad (JMO) [Steven O'Hagan]; the IMOK Olympiad papers [Andrew Jobbings]; the Intermediate Kangaroo (IK) [David Crawford and Paul Murray]; Senior Kangaroo (SK) [Carl James and David Crawford]. Those involved are listed below.

Steve Barge	(Sacred Heart Catholic College)	S
Dean Bunnell	(Queen Elizabeth GS, Wakefield)	S / IMOK / JMO
Kerry Burnham	(Torquay Boys' Grammar School)	I&J
Mike Clapper	(Australian Mathematics Trust)	I&J
James Cranch	(University of Sheffield)	IMOK
David Crawford	(Leicester Grammar School)	SK / IK / I&J
Karen Fogden	(Henry Box School, Witney)	S / I&J / JMO
Mary Teresa Fyfe	(Hutchesons' GS, Glasgow)	S / IMOK / JMO
Carol Gainlall	(Park House School, Newbury)	I&J
Tony Gardiner	(Birmingham)	I&J / IMOK / JMO
Nick Geere	(Kelly College)	S
Rachel Greenhalgh	(UKMT)	I&J
Michael Griffiths	(Warrington)	S / IMOK
Howard Groves	(ex RGS, Worcester)	S / I&J / IMOK / JMO
Jo Harbour	(Wolvercote Primary School)	JMO
Carl James	(Leicester Grammar School)	SK
Andrew Jobbings	(Arbelos, Shipley)	S / I&J / IMOK / JMO
Calum Kilgour	(St Aloysius College)	JMO
Gerry Leversha	(formerly St Paul's School)	IMOK
Paul Murray	(Lord Williams School, Thame)	I&J / JMO / IK
Steven O'Hagan	(Hutchesons' GS, Glasgow)	JMO
Andy Parkinson	(Beckfoot School, Bingley)	IMOK
Stephen Power	(St. Swithun's School, Winchester)	I&J
Peter Ransom	(Southampton)	I&J
Lionel Richard	(Hutchesons' GS, Glasgow)	S
Alan Slomson	(University of Leeds)	S / I&J
Alex Voice	(Westminster Abbey Choir School)	I&J / JMO

It is appropriate at this stage to acknowledge and thank those who helped at various stages with the moderation and checking of these papers: Adam McBride, Peter Neumann, Stephen Power, Jenny Ramsden and Chris Robson.

Summer School Staff

Summer School for Girls – August 2014

Beverley Detoeuf	Paul Fannon	Victor Flynn
Sam Ford	Bob Gray	Howard Groves
Jo Harbour	Vesna Kadelburg	Vinay Kathotia
Lizzie Kimber	Frances Kirwan	Vicky Neale
Peter Neumann	Claire Rebello	Konni Rietsch
Dan Schwarz	Alan Slomson	Geoff Smith
Sophie Wragg	Alison Zhu	

Oxford Week 1 – August 2014

Anne Andrews	Beverley Detoeuf	Michael Bradley
Sue Cubbon	Richard Earl	James Gazet
Fraser Heywood	Gerry Leversha	Martin Orr
Laura Piper	Alan Slomson	Geoff Smith
Dominic Yeo		

Oxford Week 2 – August 2014

Andrew Carlotti	Philip Coggins	David Crawford
Natasha Davey	Beverley Detoeuf	Mary Fortune
Howard Groves	Maria Holdcroft	Andrew Jobbings
Vesna Kadelburg	Sahl Khan	Frances Kirwan
Vicky Neale	Dominic Rowland	Dan Schwarz
Geoff Smith	Dominic Yeo	

Leeds Week 1 – July 2015

Robin Bhattacharyya	Michael Bradley	Oliver Feng
James Gazet	Maria Holdcroft	Gerry Leversha
Paul Russell	Charlotte Squires-Parkin	Andras Zsak

Leeds Week 2 – July 2015

Katie Chicot	Julia Collins	Mary-Teresa Fyfe
Tony Gardiner	Andrew Jobbings	Steven O'Hagan
Catherine Ramsay	Dominic Rowland	Alan Slomson
Karthik Tadinada	Dorothy Winn	

264

Patricia Andrews	Beth Ashfield (C)	Ann Ault (W)
Martin Bailey	Anne Baker	Bridget Ballantyne
Andrew Bell	Zillah Booth	Elizabeth Bull
Dean Bunnell (W)	Kerry Burnham	Keith Cadman (W)
Madeleine Copin (C)	Elaine Corr	James Cranch
David Crawford (W)	Rosie Cretney	Alex Crews
Mark Dennis	Dusty de Sainte Croix	Geoffrey Dolamore
Sue Essex (W)	Sheldon Fernandes	Jackie Fox
Roy Fraser	Helen Gauld	Peter Hall
Karl Hayward-Bradley (W)	Terry Heard	Fraser Heywood (W)
Sue Hughes	Sally Anne Huk	Pam Hunt
Andrina Inglis	Andrew Jobbings (W)	Tricia Lunel
Pat Lyden	Cara Mann	Matthew Miller
Hilary Monaghan	Steve Mulligan	Helen Mumby
Peter Neumann (W)	Pauline Noble	Martin Perkins (C)
Dennis Pinshon	Valerie Pinto	Vivian Pinto
Stephen Power	Jenny Ramsden (C)	Peter Ransom (W)
Heather Reeve	Valerie Ridgman	Syra Saddique
Nikki Snellgrove	John Slater	Alan Slomson
Graeme Spurr	Anne Strong	Penny Thompson
Ian Wiltshire	Rosie Wiltshire	

Additional local helpers and organisers at TMC host venues

Anthony Alonzi	Morag Anderson	Emma Atkins	Sharon Austin
Russell Baker	David Bedford	Helena Benzinski	Will Bird
Richard Bradshaw	Frank Bray	Nigel Brookes	David Brooks
Helen Burton	Joseph Carthew	Maxine Clapham	Kath Conway
Rebecca Cotton-Barratt		Barry Darling	Andrew Davies
Sonia Dragone	Lucy Gill	Nick Hamshaw	Laura Harvey
Gary Higham	Stephen Hope	George Kinnear	Neil Maltman
Arnita Manandhar	Helen Martin	Lin McIntosh	David McNally
Iain Mitchell	Marijke Molenaar	Heather Morgan	David Morrissey
Julie Mundy	Damian Murphy	Colin Reid	Peter Richmond
John Rimmer	John Robinson	Andrew Rogers	Amelia Rood
Ann Rush	Amanda Smallwood	Dominic Soares	Richard Stakes
Gerard Telfer	Paul Thomas	Annette Thompson	Rachel Tindal
Aaron Treagus	Sam Twinam	Danny Walker	Jo Walker
Liz Ward	Phillip Watson	Dave Widdowson	Jake Wright

STMC coordinators and regional helpers
[also involved in the writing (W) and checking (C) of materials where indicated]

Hugh Ainsley, Patricia Andrews, Ann Ault, Matthew Baker (W), Andrew Bell, Zillah Booth, Cath Brown (C), Dean Bunnell, Kerry Burnham (W), Valerie Chapman, Tony Cheslett (W), Antony Collieu (W), David Crawford (W), Rosie Cretney, Alex Crews (C), Laura Daniels, Geoffrey Dolamore, Sue Essex, Karen Fogden (W), Helen Gauld, Nick Geere, Douglas Hainline, Peter Hall (W), Mark Harwood (W), Paul Healey, Terry Heard, Fraser Heywood, Alexandra Hewitt (W), Sue Hughes, Sally Anne Huk, Pam Hunt, Andrina Inglis, John Lardner, Pat Lyden, Peter Neumann, Charlie Oakley (W), Martin Perkins (C), Dennis Pinshon (W), Lorna Piper, Stephen Power, Jenny Ramsden (C), Alexandra Randolph (W), Katie Ray (W), Heather Reeve (C), Valerie Ridgman, John Silvester (C), John Slater, Alan Slomson (C), Anne Strong, Penny Thompson, Lynne Walton, James Welham, Ian Wiltshire, Rosie Wiltshire

Maths Circles Speakers and Event Leaders

Pat Andrews, Richard Atkins, Anne Baker (L), James Beltrami, Stacey Bermey, John Berry, Dean Bunnell (L), Kerry Burnham, Luis Cereceda, Valerie Chapman, Katie Chicot, Justin Clements, Philip Coggins, Julia Collins, Madeleine Copin, James Cranch, Yvonne Croasdaile, Mark Dennis (L), Ceri Fiddes, Mary Teresa Fyfe, Tony Gardiner, Gwyn Gardiner, Nick Geere, Stuart Haring, Chris Harrison, Paul Healey, Hugh Hill, Ina Hughes, Susie Jameson-Petvin (L), Andrew Jobbings, Tom Killick, Lizzy Kimber, Cesar Lecoutre, Gerry Leversha, Kevin Lord, Tom Lyster, Emily Maw, Adam McBride, Vicky Neale, Peter Neumann, Steven O'Hagan, Stephen Power (L), Catherine Ramsey (L), Peter Ransom, Dominic Rowland (L), Ian Russell, Peter Scott, Amanda Shaw, John Slater (L), Alan Slomson (L), Geoff Smith

Susan Sturton	Karthik Tadinada	Owen Toller
Nick Turnbull	Charles Vereker	Lynne Walton
James Welham (L)	Amy Wilson	Dominic Yeo

We thank the following schools and universities for hosting Maths Circles events

Carmel College, St Helens	Conyers School, Yarm
Harrogate Ladies' College, Harrogate	Hutchensons' Grammar School
Lincoln Minster School, Lincoln	North Chadderton High Sch., Oldham
Oundle School, Oundle	St Paul's School, London
St Swithun's School, Winchester	Wells Cathedral School, Somerset
Wycliffe College, Gloucestershire	

BMOS Mentoring Schemes
James Cranch (Director)

Junior Scheme Coordinator: John Slater

Intermediate external mentors:

Alice Ahn	Sally Anne Bennett	Andrew Jobbings
Ella Kaye	Zoe Kelly	Roger Kilby
Radhika Mistry	Gordon Montgomery	Robbie Peck
David Phillips	Oliver Sieweke	Ian Slater
Alan Slomson	Hugo Strauss	Alasdair Thorley
George Welsman		

Senior Scheme Coordinator: Andre Rzym

Senior external mentors:

Anne Andrews	Andrea Antoniazzi	Sam Banks
Katriona Barr	Jamie Beacom	Philip Beckett
Natalie Behague	Alasdair Benjamin	Don Berry
Ruth Carling	Nicholas Chee	Andrea Chlebikova
Xenatasha Cologne-Brookes	Samuel Crew	John Cullen
Pawel Czerniawski	Natasha Davey	Chris Ellingham
Robin Elliott	Oliver Feng	John Fernley
Mary Teresa Fyfe	Simon Game	James Gazet
Gabriel Gendler	Julian Gilbey	Esteban Gomezllata Marmolejo
James Hall	Matthew Haughton	Paul Healey
Adrian Hemery	Fraser Heywood	Edward Hinton
Maria Holdcroft	Daniel Hu	Ina Hughes
Mihail Hurmuzov	Michael Illing	Susie Jameson-Petvin

Sahl Khan
Jonathan Lee
Daniel Low
David Marti-Pete
Pavlena Nenova
Ramsay Pyper
Julia Robson
Ben Spells
Federica Vian
Perry Wang
Catherine Wilkins
Paul Withers
Fabian Ying

Mark Knapton
Gerry Leversha
Chris Luke
Gareth McCaughan
Peter Neumann
Katya Richards
Roberto Rubio
Stephen Tate
Paul Voutier
Kasia Warburton
Daniel Wilson
Dominic Yeo

Robert Lasenby
Michael Lipton
Matei Mandache
Vicky Neale
Keith Porteous
Jerome Ripp
Jack Shotton
Oliver Thomas
Paul Walter
Mark Wildon
Dorothy Winn
Michael Yiasemides

Advanced Scheme Coordinator: Richard Freeland

Advanced external mentors:

James Aaronson
Adam Goucher
Henry Liu

Andrew Carlotti
Tim Hennock
Jordan Millar

Richard Freeland
Freddie Illingworth
Joseph Myers

UKMT Publications

The books published by the UK Mathematics Trust are grouped into series.

The *YEARBOOKS* series documents all the UKMT activities, including details of all the challenge papers and solutions, lists of high scorers, accounts of the IMO and Olympiad training camps, and other information about the Trust's work during each year.

1. 2014-2015 Yearbook

This is our 17th Yearbook, having published one a year since 1998-1999. Edited by Bill Richardson, the Yearbook documents all the UKMT activities from that particular year. They include all the challenge papers and solutions at every level; list of high scorers; tales from the IMO and Olympiad training camps; details of the UKMT's other activities; and a round-up of global mathematical associations.

Previous Yearbooks are available to purchase. Please contact the UKMT for further details.

PAST PAPERS

1. *Ten Years of Mathematical Challenges 1997 to 2006*

Edited by Bill Richardson, this book was published to celebrate the tenth anniversary of the founding of UKMT. This 188-page book contains question papers and solutions for nine Senior Challenges, ten Intermediate Challenges, and ten Junior Challenges.

2. *Past Paper Booklets and electronic pdfs – Junior, Intermediate, Senior Challenges and follow-on rounds.*

We sell Junior, Intermediate and Senior past paper booklets and electronic pdfs. These contain the Mathematics Challenge question papers, solutions, and a summary chart of all the answers.

The JMO booklet contains four years' papers and solutions for the Junior Mathematical Olympiad, the follow up to the JMC.

The 2013 IMOK booklet contains the papers and solutions for the suite of Intermediate follow-on rounds – the Grey Kangaroo, the Pink Kangaroo, Cayley, Hamilton and Maclaurin. Electronic versions of the IMOK Olympiad rounds and the JMO are also available.

BMO booklets containing material for the British Mathematical Olympiad Round 1 or 2 are also available.

The *HANDBOOK* series is aimed particularly at students at secondary school who are interested in acquiring the knowledge and skills which are useful for tackling challenging problems, such as those posed in the competitions administered by the UKMT.

1. *Plane Euclidean Geometry: Theory and Problems,*
   AD Gardiner and CJ Bradley

An excellent book for students aged 15-18 and teachers who want to learn how to solve problems in elementary Euclidean geometry. The book follows the development of Euclid; contents include Pythagoras, trigonometry, circle theorems, and Ceva and Menelaus. The book contains hundreds of problems, many with hints and solutions.

2. *Introduction to Inequalities*, CJ Bradley

Introduction to Inequalities is a thoroughly revised and extended edition of a book which was initially published as part of the composite volume 'Introductions to Number Theory and Inequalities'. This accessible text aims to show students how to select and apply the correct sort of inequality to solve a given problem.

3. *A Mathematical Olympiad Primer*, Geoff C Smith

This UKMT publication provides an excellent guide for young mathematicians preparing for competitions such as the British Mathematical Olympiad. The book has recently been updated and extended and contains theory including algebra, combinatorics and geometry, and BMO1 problems and solutions from 1996 onwards.

4. *Introduction to Number Theory*, CJ Bradley

This book for students aged 15 upwards aims to show how to tackle the sort of problems on number theory which are set in mathematics competitions. Topics include primes and divisibility, congruence arithmetic and the representation of real numbers by decimals.

5. *A Problem Solver's Handbook*, Andrew Jobbings

This recently published book is an informal guide to Intermediate Olympiads, not only for potential candidates, but for anyone wishing to tackle more challenging problems. The discussions of sample questions aim to show how to attack a problem which may be quite unlike anything seen before.

The *EXCURSIONS IN MATHEMATICS* series consists of monographs which focus on a particular topic of interest and investigate it in some detail, using a wide range of ideas and techniques. They are aimed at high school students, undergraduates, and others who are prepared to pursue a subject in some depth, but do not require specialised knowledge.

1. *The Backbone of Pascal's Triangle*, Martin Griffiths

Everything covered in this book is connected to the sequence of numbers: 2, 6, 20, 70, 252, 924, 3432, ... Some readers might recognize this list straight away, while others will not have seen it before. Either way, students and teachers alike may well be astounded at both the variety and the depth of mathematical ideas that it can lead to.

2. *A Prime Puzzle*, Martin Griffiths

The prime numbers 2, 3, 5, 7, ... are the building blocks of our number system. Under certain conditions, any arithmetic progression of positive integers contains infinitely many primes, as proved by Gustave Dirichlet. This book seeks to provide a complete proof which is accessible to school students possessing post-16 mathematical knowledge. All the techniques needed are carefully developed and explained.

The *PATHWAYS* series aims to provide classroom teaching material for use in secondary school. Each title develops a subject in more depth and detail than is normally required by public examinations or national curricula.

1. *Crossing the Bridge*, Gerry Leversha

This book provides a course on geometry for use in the classroom, re-emphasising some traditional features of geometrical education. The bulk of the text is devoted to carefully constructed exercises for classroom discussion or individual study. It is suitable for students aged 13 and upwards.

2. *The Geometry of the Triangle*, Gerry Leversha

The basic geometry of the triangle is widely known, but readers of this book will find that there are many more delights to discover. The book is full of stimulating results and careful exposition, and thus forms a trustworthy guide. Recommended for ages 16+.

The *PROBLEMS* series consists of collections of high-quality and original problems of Olympiad standard.

1. *New Problems in Euclidean Geometry*, David Monk

This book should appeal to anyone aged 16+ who enjoys solving the kind of challenging and attractive geometry problems that have virtually vanished from the school curriculum, but which still play a central role in national and international mathematics competitions. It is a treasure trove of wonderful geometrical problems, with hints for their solutions.

We also sell:

1. *The First 25 Years of the Superbrain*, Diarmuid Early & Des MacHale

This is an extraordinary collection of mathematical problems laced with some puzzles. This book will be of interest to those preparing for senior Olympiad examinations, to teachers of mathematics, and to all those who enjoy solving problems in mathematics.

2. *The Algebra of Geometry*, Christopher J Bradley

In the 19th century, the algebra of the plane was part of the armoury of every serious mathematician. In recent times the major fronts of research mathematics have moved elsewhere. However, those skills and methods are alive and well, and can be found in this book. The Algebra of Geometry deserves a place on the shelf of every enthusiast for Euclidean Geometry, amateur or professional, and is certainly valuable reading for students wishing to compete in senior Mathematical Olympiads. For age 16+ mathematicians.

3. The UKMT is the European agent for a large number of books published by the Art of Problem Solving (http://www.artofproblemsolving.com/).

To find out more about these publications and to order copies, please go to the UKMT website at www.publications.ukmt.org.uk.

In addition to the books above, UKMT continues to publish its termly Newsletter, giving the latest news from the Trust, mathematical articles, examples from Challenge papers and occasional posters for the classroom wall. This is sent free to all schools participating in the UKMT Maths Challenges.